The Reality of Building A Horse Barn and Farm

By
Janet G. Miller

ISBN 978-0-557-57619-7

http://www.americanrealitybooks.com/

Contents

Prologue: : Page 4

Preface : . . . . : Page 5

Chapter One : Introduction : Page 7

Chapter Two : Before You Purchase the Land : Page 13

Chapter Three : After You Purchase the Land – Site Development : Page 58

Chapter Four : Horse Barn Planning – Builders and Banks : Page 97

Chapter Five : Hay Barn Planning, Commercial Facilities, Waste Disposal : Page 115

Chapter Six : Horse Barn Planning – Second Phase : Page 139

Chapter Seven : Arena Design and Construction : Page 212

Chapter Eight : Finishing the Horse Barn Interior : Page 237

Chapter Nine : Farm Equipment and Final Landscaping : Page 261

Chapter Ten : Epilogue : Page 276

Addendums: Poem : Page 282

Southwestern United States : Page 283

Lessons : Page 293

Photographs : Page 296

Diagrams : Page 302

Index : Page 303

PROLOGUE

It starts with a simple drawing of a couple of rectangular shapes on a 3 x 5 card.

Don't be fooled!

What looks simple turned into this 135,236 word document to support that little 3 x 5 card drawing.

When it comes to horse barns, design is not enough.

Nor is it enough to just be an equestrian.

No, the reality is that when building a horse farm, it is essential that the design and construction of the farm support the operation of the farm.

That means horse barn building requires the insight from a combination of horse knowledge, barn building knowledge, and horse farm operation knowledge.

That combined knowledge results in the "reality" of barn planning!

If you're planning to build a horse barn/farm, you MUST first read this book!

PREFACE

My husband and I were not new to building houses nor were we new to owning horses when we began our horse farm project. We had directed the construction of two custom houses from scratch—one in Connecticut and one in west Michigan—we had directed the completion of two spec houses—one in New Jersey and one in Ohio—and lived in two other new houses—one in Pennsylvania and one in Massachusetts. My children and I had owned our own horses and had been heavily involved in horse showing for about 16 years before we began the construction of our own horse farm, which we planned to initially be a private facility and then convert into a commercial facility. We were more prepared and knowledgeable at the beginning of the horse farm project than we had ever been with any of our previous home building projects. Yet nothing could have prepared us for the reality of what happened during our journey to make our horse farm dreams come true. The level of dishonesty, incompetency, corruption, and lack of ethics that we encountered were not only shocking to us, but frankly, were very disheartening. There were also several things that we just didn't know and took us by surprise.

What we couldn't do, however, once the project began, was walk away from it. There are huge upfront outlays of money involved with building a horse farm from scratch that are difficult to impossible to recover. If we had known the information that I am about to share with you in this book before we started our project, I am sure the project would have been completed with a reduction in time and cost and with better results.

We found the process of building a horse farm to be very confusing and challenging. But the fact is that as the architect, design, financial and construction industry currently exists, there is no straightforward way to get anything done. As well prepared as we thought we were to build a horse farm, we were repeatedly foiled by the actions of others that made it impossible for us to predict any outcomes from our actions because there was no way to be able to predict dishonesty, bankruptcy and acts of corruption by others. Then, as I stated, once we were into the project, and tens of thousands of dollars had been spent and were basically unrecoverable, there was no way out of the project that would protect our investment other than to complete it.

There are lots of "how to" books that exist that give a superficial glance at an order of go in how a horse farm or barn extension to an existing property should be built that make you think that is the straight forward way things will occur. However, the reality of the situation does not even come close to anything that could be categorized as being simple or straight forward. Even though I personally experienced the entire process, I still found it confusing and difficult to be able to write it all down in a book and try to make sense out of it. In talking to others who have dealt with horse farm and barn construction, we know that our real experience is not unique—nevertheless the experience remains shockingly eye opening to those who have not yet personally experienced it.

DISCLAIMER: As the reader of this book, you must define your own objectives and your own plan for building your own horse farm. You must accept the liability and the responsibility for creating your plan, and the consequences of your plan. Although some of the language in this book might sound like directions for you, your architect, designer or your builder, this book is not intended to give you, or anyone, specific directions for creating that plan or building either your farm or your barn. This book is intended only to share information that we learned which you can either accept or reject, and apply with or without modification. And, please understand that everything written, expressed or implied in this book is written, expressed or implied as merely my opinion.

In writing this book, I felt that the best approach I could take was to offer you both real stories based on our real experiences to clarify and emphasize certain issues, and then include important hard facts and information that I felt you needed to know in order to try and avoid the many pitfalls that exist in the barn building industry that we learned about from our experience. I have presented the information roughly in the order that it occurred in our real life situation. I have also given numbered insights which I have labeled as "lessons" in the writing that are points at which my husband and I had epiphanies (profound insights we learned) that we felt were important to share with you, and important enough to emphasize by highlighting them and numbering them.

Chapter One

Introduction

Photo #1: Lush Green Pastures at the End of a Late Summer Day

Standing there in the middle of all those acres, at the end of a long, lazy, late summer day, the gentle early evening breezes passing over and around you, as you watch the big, bright, orange ball of summer sun slowly sinking from the western sky, the horses grazing peacefully among the long, lush grasses of the pastures, you just can't help but think to yourself, "What could be more beautiful and more restorative to the soul than this?"

And the answer is … nothing.

And, so, you decide, right then and there, that some day, you will stop boarding your horses at someone else's facility, and bring your horses home to live with you on your own property. And that means you will eventually be building or buying your own horse farm.

After all, what could be simpler? Aren't you already totally knowledgeable about horse farms and their operation from your current boarding at a farm? Aren't you basically already taking care of your own horse or horses at the boarding facility? Haven't you learned just about all the information you would need to know at this point in your horsemanship in order to go off on your own? Aren't you at least as capable and intelligent as anyone you see running anything at the farm where you are now boarding? So why not have your own place? Why not stop paying someone else all that money when you could do it all yourself? Why not own and run your own farm and have your horses right outside your door and ready to ride?

That all sounds rational. That all sounds plausible. In fact, that all sounds downright happy and practical especially from a financial point of view.

So why not do it?

There are plenty of books and other references out there too. You could do a lot of reading, and thinking, and planning beginning right now. Then come the day when you're "ready" to build your own place, you'll feel sufficiently knowledgeable and ready to go.

Also sounds rational.

In fact, that's exactly how my mind thought at one point in my life about four years before I began writing this book, when I had about sixteen years of riding and showing under my breeches, so to speak. And I still love to stand in the middle of green acres, feel the breeze, watch the sunset, and hopefully watch horses that are peacefully grazing. But at this point, I'm also a lot of other things that are not necessarily so romantic. What I am, however, is a lot smarter and wiser than I was seven years ago.

And that's why I'm writing this book.

I want to give you horse farm dreamers a more realistic grasp of what's involved in building and having your own farm. I also, hopefully, want to help you avoid stepping into some of the potholes I did to make your experience an overall easier one than mine. I should also mention that I have written this book assuming that you will be trying to build your farm on a limited budget.

Nevertheless, this book is also important for those of you who have unlimited dreams and the money to match it. The point being that even if you're a billionaire building a farm, and you only plan to interact with an expensive architect you've hired for the project, you will still find that the architect is going to ask you a thousand questions about what you want on your farm. And, if you don't have the answers to the architect's questions, then the architect is going to do things his way. And, if the architect does it his way, then that might not necessarily be in your or your horses' or your pocketbook's best interest, and potentially make you a very disappointed farm owner in the end. Then too, there are architects who will refuse to answer the questions for you because they will not assume the liability should their choices not turn out the way you wanted them once you "see" them done—such result mismatches can also potentially cause your project to repeatedly stall. So reading this book will, at a minimum, make you savvy about the kinds of questions you will be asked about building a farm when you do meet with your architect or perhaps inspire you to do some research so you know what you really want.

This book will also help those of you who might be considering purchasing a horse farm that someone else built to see if the existing facility meets your standards before you put your money into it. It is also a good read for those of you who might just want to have some insight into what it's like to build a horse farm and live on one.

And least I forget my "kinsmen" of so many years, those of you who ride and board, this book is for you as well because it sets the standard against which you can compare what you are actually getting for your money at your boarding facility, and if your horse is really living at The Ritz--like your barn owner is claiming he or she is--and getting the best possible care. Hopefully, at a minimum, reading this book will inspire you to demand higher standards at your boarding facility or, even better, will inspire you and a few friends to build a shareholder, coop farm so that you can get out from under the control of a trainer and finally be in charge of your own life again, and that of your horse.

I have written the book in a way that lets you experience my personal journey into buying land, planning our farm, building it, and then moving into the farm and living on it. The book is based on a horse farm that was built to exist in a four season climate and has a basic pole barn structure with its accompanying metal exterior walls. Whereas you may be thinking about building a wood barn or a brick barn or perhaps an open barn as is typical in the Southwest, it is important to remember that the inside of barns and their planning share a commonality in design which you will learn about in this book. The exterior of the barn is basically just the shell of the barn, and mainly involves just a decision as to whether the exterior will be wood, metal, brick, or open — each type of exterior material having its commensurate supporting structure associated with it and a roof. The point being that horse barns are far more than their exterior walls and roofs.

Yes, there are lots of books on the market as to "how" to build a barn, "how" to handle a horse, and the like, but then there's the reality of integrating the design with the construction so that it's a livable and workable horse farm. There are all kinds of things that will happen when you are actually living on your own horse farm that you never would have thought about in your wildest dreams. These things can, and surely will, happen in one form or another, and they are directly related to the design and construction of a horse farm. That's what this book is about.

An additional benefit to reading this book is that I can share with you several of the horsemanship experiences I have had over more than 20 years of my owning, riding, and showing horses and how these experiences related to the design, construction, and operation of our horse farm. Writing this book about building a horse farm has given me a venue in which to share with you a wealth of information that I would not ordinarily be able to share. Any gain in horsemanship knowledge is always better for both you and your horse's health and happiness. Reading this book will give those of you who are new to riding or those of you who are just a few years into it, a whole lot more horse sense in a short period of time than can ordinarily be garnered. Remember that as a general rule, the average person is "in" to horses, and then "out" of horses within a 7 year period!

I should note that everything I did in planning and building my farm was from a horse's and horsemanship point of view based on my many years of personal experience. Once you read the book, you will understand why it is not sufficient to consult with "just" an architect or a designer or "just" one type of builder when building a horse farm. You will also be able to assess the value and functionality of any horse facility you might be considering purchasing or one in which you are currently boarding or considering boarding your horses. In other words, you have to know horses and horsemanship in order to build or purchase a farm or board in a barn that is happy, safe, and healthy for you and your horses.

Please remember that all horse behavior observations contained in this book are just that—observations that I have made with my real horses, on my real farm. I am not a veterinarian, nor am I a university professor or university researcher with published scientific studies. I am just a real horse and farm owner/operator dealing with real life observations and situations concerning horses on my farm. Four years ago I was just like 98% of you who are reading this book and craved real world insight into what it's like to be with horses 24/7. I was continually frustrated by obtuse "opinions" (though potentially based on "scientific" research) offered in fragments of information, gained from periodic, distanced, objective observation (which means no 24/7 interaction with horses) of horses in order to earn a Ph.D. in animal behavior or to acquire a research grant to study animals.

True, this book is written from my own limited and very personal experiences. Also true is the fact that I can't allow for all the variables you will have to consider and that you will encounter on your personal project. Please understand that everything contained herein is only offered to serve as eye openers to "jolt" your brains into a constant state of awareness and analytical thinking while planning and building your own farm.

The book is intended to be a conceptual—not a literal—step by step guide based on how we built our horse farm and our experiences. It is meant as a general guide on when, how, what, and why to think, and then what to do after you've done your thinking! By sharing this with you, you will hopefully get insights into those little things people seem to somehow consistently forget to mention to you about horses, horse handling, and horse structures that make all the difference in the happiness level of the final result.

Case and point: A really nice man came over to our barn one day to fix one of our automatic waterers. He told us that he knew exactly where our farm was because he and his wife had regularly visited the farm while under construction. He told us that he had also recently built a barn, and it was clear as he spoke about his barn that he had a great deal of pride about his final result. However, as he stood in our barn and chatted with us about our barn, he repeatedly noted this or that we had included in our barn that he had needed to put in as an "add on" in his barn after his barn was "done." I smiled to myself thinking how much he would have benefitted from reading this book, but of course I didn't say anything. And the fact is that it's true that you can "add on" a lot of things. Nevertheless, what is also true is that each add on will look like an add-on, and function like an add-on—a smooth, integrated functionality will never be achieved without advance planning. There are also some fundamental things that you cannot realistically change about your barn or farm once it is built—or at least they are things that cannot be done without a great deal of difficulty, disruption, and expense. For example, it would be impractical to install larger, higher quality windows after the fact. Meanwhile, back to the story, the saddest expression came across this very nice man's face when I mentioned our stall windows. His self absorbed, focused eyes left our conversation at the mention of the stall windows, as he looked up from the waterer he was repairing to the window in the stall. He stopped all motion for a second. Then in a quiet little voice filled with notable disappointment he shared with us that he "hadn't thought to put windows in his horses' stalls."

Building a farm is actually very similar to parenthood. There comes a point in your life when you are sure that becoming a parent all makes sense. You feel like you've got a good handle on what parenting is all about, you've seen others doing it, and "handling" it all, and so, why not you? You've read plenty of books about parenting, attended classes and demonstrations. So what could possibly happen to take you by surprise when you become a parent? And yet there isn't a parent out there who couldn't answer that question with at least 50 little stories of "surprise" and "horror" that presented themselves in the reality of the parenting process.

Please remember that everything in the book is "in my opinion." That's especially important for those of you in the architectural, design or construction business who might find my honesty too harsh for your liking. Also please note that any pricing offered is a "ball park figure" in that prices are constantly changing.

Also, please forgive me for using the "I" word over and over again. I use it only to provide information from my personal experiences for your benefit. Please also don't ever feel that I think I have a perfect farm or that everything I thought and did was perfect. In fact, the reality is that after all the compromises I made during my planning and construction that deviated from my original dream barn, I was flat out shocked at how happy and functional the barn turned out to be! I was literally shocked! The fact is that sharing my experiences to help you in the planning and construction of your farm does necessarily become rather myopic. So, please, again, forgive me for overuse of the first person in the book!

So here's the "real deal" on a lot of things to be on the lookout for on the days when your cherished ideals start smacking into the harsh walls of reality.

Chapter Two

Before You Purchase the Land

First, let me officially welcome you to the wonderful, wacky world of building a horse farm! I sincerely hope that you enjoy reading about my adventures and learn reams of important information from them so as to avoid the many mistakes that frequently happen—including the ones that I made—in the barn building process. May I also suggest that on the day you actually start your own project that you take a good look at yourself in the mirror and take an assessment of who you are on that day and how you feel about horses and riding, how you define the word "equestrian," and how you define yourself in regard to being an "equestrian." Why? Because I'm pretty sure that by the time you finish your project, and then move in and actually manage your own horse farm, that you will not be the same person you were on the day you started, nor feel the same way about horses or riding, nor define the word equestrian in the same way as you now do. Is that a negative statement about building a barn? Well, maybe, and maybe not. That will be something for you to determine, and then answer for yourself, when you are done!

So now that we have been officially introduced to each other, how about if we enter our magical journey into this funny and frustrating adventure by talking about **land**? That seems to be a good place for us to start, doesn't it?

After all, what could be simpler than land? God already made it. It exists. There's plenty of it. Just take a ride out of the urban or suburban areas and there it is—land, land, and more land. Beautiful land, rolling land, flat land, treed land, farm land—you name it, you can find it. Plus, there are even plenty of "for sale" signs on land as you pass by. So surely there's a piece of land out there just right for you and your horse farm, isn't there?

Well, maybe, and, then again, maybe not.

The first question you should ask yourself about land is "**who owns the land**?" Most likely you will answer that some person or organization holding the deed to the land obviously owns the land; and you would be right. Then ask yourself "**who can sell the land**?" The obvious answer would be the person who holds the deed to the land can sell the land, and that would also be correct. That was simple, wasn't it?

But now try and find a farmer or deed holder on your own or with a realtor representing you who is willing to sell you even a small parcel of their land. Not likely, yet also not impossible. What you might find is that the best way to buy land is to look beyond what is listed by residential realtors. We found our land by talking to **commercial realtors** who specialized in land sales and development.

But let's assume that you have found a parcel of land that's available for you to purchase. Now ask yourself **what you, as the buyer, will be "free" to do with the land** once you've purchased it. And, based on the preceding logic flow, you would most probably answer that you can do anything you want to do with the land.

And that answer would be **wrong**.

But it's not only wrong in the more obvious sense of not being able to build a horse farm in the middle of a suburban development or an industrial park. No, because it could also be wrong even if you wanted to build a farm in the middle of farmland.

Well, that's ridiculous you say. Why couldn't you build a horse farm in the middle of miles and miles of farmland?

It has to do with **zoning**.

As an example, my husband and I called the realtors listed on the "for sale" sign on a beautiful piece of land situated squarely within miles and miles of farmland that a friend had called to tell us was available knowing that we wanted to build a horse farm. We were particularly excited about this piece of land because the people from whom we had bought our elder Jack Russell lived right next door to the land which meant that we would have our friends as neighbors. So we arranged to meet with the realtors on the property at 10 a.m. the next day.

The next day, we pulled into the "driveway" of the vacant land and stopped right off the busy roadway, at the head of the entrance to the property, where there was one of those sixteen foot wide, metal tube farm gates across the gravel driveway, suspended between two, heavy duty, wooden posts. The gate was tied shut with a knotted rope and clearly sported a "no trespassing" sign. The realtor arrived almost immediately after us, pulling up to a stop in front of the closed gate. He popped out of his car with one of those big, realtor type smiles meant to convey unending, open honesty, and offered to unknot the rope telling us that there wasn't any problem "trespassing" in order to view the property.

While the realtor untied the rope, we asked several questions including how many acres were for sale (80 of them, divisible into either 20/20/40 or 40/40), and who owned the land, and why they were selling the land. The realtor told us that a doctor had bought the land for his wife to build a dressage facility, but the doctor had suddenly been transferred to another state and was therefore selling the property. That all sounded logical and inviting, especially considering someone else had seen this very piece of land as a good potential for a horse farm. We all got back into our cars and we repositioned our car so that the realtor could lead the way onto the land. Only the front of the land could be seen from the roadway in that it went uphill from the main road. As we went up the hill, we were excited to note that there was a row of huge old trees along the graveled driveway. Our imaginations ran wild with how lovely the trees would be once we put in our permanent drive leading back to our house and barns and how these trees must have been planted many years ago by someone who had happily lived here before. We expected to see nothing but the natural land as we drove along, and at this point along the road we couldn't see anything other than grass, land, and trees. The driveway was still going gradually uphill, as was our excitement level, when after a good five acres off the main road, the driveway and the land finally began to fully flatten out. As we drove on we passed through a cross section of mature trees that blocked a clear view back to the next five or so acres of land before there was another thick stand of woods that populated the remaining 20+ rear acreage.

Shock. Between those back, two wooded areas, way back from the main road, almost hidden in plain sight by its location, was a pasture shaped area with about 200 feet of wood perimeter fencing, a frost proof water pump over a well, and a very small, totally abandoned, barn like structure with giant spool hay bales dumped inside the structure that blocked all the doors. It was all very eerie. We were immediately grasped by the feeling that something was very wrong here.

What was before us just didn't sit right. There had definitely been something going on back here--something very incomplete and sketchy. It was also something that certainly didn't fit in with a story about a doctor's wife and a big dressage facility. In fact, it didn't seem to fit in with anything above board. After all, who would have any kind of animal living back here so isolated and minimally—most especially horses? Why would someone do such a thing? What would necessitate such a situation? We certainly didn't know. So we asked the realtor what was up. The realtor in turn said he had no idea either, and that he was as surprised as we were to find any structures back here. He said that he had understood that this was "just land." He explained that he and his wife had just gotten this listing and that they had never actually been on the land before. But he said that he would call his wife back at the office and try and find out what was up with the fencing, well, pump, and building.

That all sounded reasonable and cooperative and we were very interested in hearing what he could find out. After all, the natural piece of land was just beautiful, and the seclusion of the back area was very peaceful—at least it would be without the "creepy" feel of the abandoned area.

We got back in our car and followed the realtor back out to the entrance barrier as he was busily talking on his cell phone, apparently to his wife. When we reached the entrance gate, he stopped and popped out of his car and came over to talk to us at our car. He told us that his wife tried calling the doctor, but he wasn't available at the moment. He also noted to us that in talking to his wife, he wasn't really sure as to what land was available or really anything about the land, and, therefore, he felt unable to represent the land to us at the moment. However, he said he would check things out and get back to us.

Well, my husband and I were not impressed with the information the realtor had now collected and shared with us. In fact we felt very uncomfortable about something he was clearly <u>not</u> saying to us and his anxiousness about it. So, we decided to go to the administration building for the township in which the land was situated and ask them about the land.

<u>Surprise!</u>

Here is what we learned. Yes, the township told us, it would be okay for us to build a horse barn and have horses on the property. Okay, we thought, great, that was good news then, right? We could consider buying the land. Yet the air was noticeably still in the room after those facts were stated. Worse, it was that kind of stillness in the air that made you feel very uneasy as if you'd done something really dumb and they weren't sure whether or not they should tell you about it.

So as we looked at their "official" faces and "saw" the "however" word written all across them, we stared back into their faces with the word "and" written across our faces, as all the while the air continued to stand uncomfortably still around us.

Well, they finally went on, there is another factor to consider.

Yes? What would that be? we asked.

Well, they said, you have to have your primary residence on the property in order to have a barn with animals in it.

Well, that's not a problem. That's exactly what we intended. Who wouldn't want to do it that way? Why did they even mention that to us? Of course we would want to have our primary residence next to the barn, we naively proclaimed. Otherwise how would we take good care of our animals?

<u>Shocker #1:</u> Well, in this township, in order to preserve the community as agrarian, they informed us that there were constraints on the use of land. Beyond Rule #1--that you couldn't build a barn without there being a primary residence built on the same piece of land--was Rule #2—that you couldn't build a house if there was another house within 240 acres of the place where you wanted to build your house.

Bingo! Zoning Problem! Considering that our friends were living in their farmhouse about five acres to the left of the center of the piece of land we were considering buying, we obviously didn't meet the minimum 240 acre separation of houses qualification making this piece of land a "no go".

So what about that little area of fencing, the pump, the well, and the little barn that was already on the land and seemed to indicate there had been some type of livestock kept there for some period of time?

Considering how far back off the road and essentially invisible that fenced area was from the road, it seemed that the area was serving as temporary housing for the dressage horses of the doctor's wife while they decided what in the world they were going to do with this very expensive piece of crop land they had been snookered into buying for their horse farm. But apparently the one thing they had figured out was that they intended to screw someone else should they be so naïve, as they obviously had been, to think they could build a horse farm on that piece of land.

NOTE: Well, as my mother used to say, "never say never." Three years after looking at that unbuildable piece of land, you'll never guess what I saw when I recently drove past the property. There was an elaborate drive and a fancy sign advertising a new subdivision of homes going in on the very same land no one was allowed to build on three years ago. So, suddenly, not only was it possible to build one home on the land even though it didn't meet the 240 acre separation criteria, it was now possible to build several homes. Can you spell payoffs? Can you spell corrupt local zoning board?

LESSON #1: Before you buy a piece of land, go to the township or county administration building or government agency responsible for zoning in that locality and find out what the building and use restrictions and rights are for the land and particularly what the restrictions and rights are for essential things such as access to the parcel including access roads and driveways, easements for access and utilities, water rights including wells, mineral rights (who has the right to do mining on or under your land), availability of electricity, septic systems, architectural and other restrictions on constructing houses and agricultural buildings including barns and arenas, how much of the land has to remain open (or undeveloped as is the case for wetlands) or put into pastures, whether there is a home owners association (HOA) or equivalent with covenants that would prevent things like disposal of manure on the land or irrigation or fertilization, and the maximum number of horses you can have on the land. This includes setback lines, separation distances between buildings, separation distances between wells and septic systems. In other words, ALL restrictions—and get them in writing. After you complete your initial investigation

and are seriously interested in buying a specific property, then it is essential you find an attorney with real estate expertise to hire, and direct your attorney in writing to do a due diligence on all the restrictions including possible title problems such as expiration of easements and violations of zoning ordinances regarding the property. In addition, find out if the restrictions can be changed later, who can change the restrictions over your objections as a property owner, and what process is required (such as application for a zoning variance), and then how the changes could affect the future viability of your farm like having a shopping center being built next to your acreage or having a new road or neighbor's driveway run through your acreage. Once your attorney has given you the green light that you can build what you want on the land and use it in the manner you want, then you can begin your planning. It might be to your advantage to consider having your attorney write a custom sales contract to protect you.

Then in our search for land, we found this nice 40 acre patch behind an old Victorian house that the owners had decided to sell off for a little income. This realtor handed us a very elaborate, professionally drawn, plot plan showing potential divisions of the property creating the impression of approval of the building sites by the township. Even though seemingly approved for building, there were still some issues with the land, like the fact that a major interstate lay at the very back edge of the property. Nevertheless, it was close to town on the front side, and had a pretty, gentle roll to the land. We decided to walk the entire piece of land with the realtor, and although we weren't overly thrilled about the land, we were curious. Upon closer questioning, we eventually asked the realtor if the land perked.

NOTE: For those of you who may be total "city" people, you should realize that public sewer and water generally does not reach out into the countryside. Your potential farm land will most likely require that you be self sufficient with both your source of drinking water (**a well with adequate quality and flow rate**) and your disposal of human waste (**a septic system**).

NOTE: **"Perked" land** is land that has been tested by the County Health Department and issued a certificate of approval by the Health Department allowing a septic system to be installed on the property. When the land has "perked," it means that the composition of the soil where the septic drain field will be located can accommodate the need of a septic system to discharge the liquid from its storage tanks via pipes to the septic drain field and allow the liquid to "percolate" down through the soil of the drain field. This percolation process of the fouled liquid through the soil allows a natural filtration process of the fouled water to take place making it ultimately "safe" to naturally distribute it back into the environment once it reaches the bottom of the septic field. There is also evaporation of some discharged liquid that occurs from the top of the field. Overall, in order for the system to work, the soil of the field must always have air pockets in it to allow flow of liquids through the soils. Therefore, absolutely

NOTHING can be built on the septic field, nor can anything of weight be constantly driven over or walked across the septic field (including horses). In other words, there can be no compression of the septic field soils.

NOTE: Just because the land perks and can have a septic system, it does NOT mean that you can put any number of bathrooms in your structures that you want. The **number of bathrooms** you can have will depend upon the size of the septic drainage field that has been approved. For example, if you want a 5 bathroom house plus an additional bathroom in the barn, making a total of 6 bathrooms on the property, and your septic system has only been approved for 3 bathrooms, then 3 bathrooms will be all that are allowed.

NOTE: Our septic field could only be dug and located on our land on the specific area the county Health Department had approved and was indicated on an official survey of our property. So the house (or other structures) that would be using the septic field had to be either: (a) built right next to the drainage field for a gravity feed of the discharge from the house into the storage tanks and then from the storage tanks into the septic drainage field; or (b) built away from the septic field which would then require an engineered septic system to pump the discharge from the storage tanks out to the septic drainage field.

Asking if the land has perked is a very important question to ALWAYS ask. In this case with this land we were considering, an answer was NOT straight forward in coming.

Well, the realtor said to us with great authority (meant to intimidate), apparently the land did perk, because this was an engineered survey she was holding in her hand, and there were indications on the document that there were areas that perked.

Well, we asked her, where exactly were those perk areas specifically indicated on the document?

Well, it says something here, she noted on one sectioned piece of land.

We then took the time to look closely at the division she had pointed to, as well as the entire survey she had in hand, and noted to her that although the one parcel did appear to have perked and been notated as such, that the other pieces of land, divided as potential building sites, lacked such "perk" notation on the document. So, we asked her about those other pieces of land and if they perked or not. We noted to her that as far as we could see, the survey said nothing about those pieces of land perking at all.

Taking another look at the survey, the realtor suddenly begged off, and said she didn't know for sure what perked. She said, however, that we should feel free to call the engineering surveyor firm and discuss the survey with them. But then she also mentioned that she felt certain that, regardless of what the engineering firm said, that we could at least do a sand mound septic field if nothing else.

A **sand mound septic system**? Why did she mention that? Sand mound septic systems aren't generally done, and aren't generally understood. Had she heard someone mention that type of system in regard to this land? Regardless, we knew that you don't go there. People don't want to have a sand mound septic system. We knew that from living in other states where they were allowed. We knew they were limited in their effectiveness at best. They were certainly something that would be more than inconvenient when serving a horse farm. As we pressed the perking issue, the realtor got more and more elusive and defensive. It became clear to us that the reason none of the plots had sold was that none of them perked.

NOTE: This acreage behind the Victorian house still lays undeveloped five years after we looked at it. There is a section on which someone has put an RV "home," which is a way "around" having a septic system so you can "live" on the property—that is if the zoning "allows" RV's to be lived in on the land which is generally unlikely. But, clearly, the land did NOT perk, and someone got "stuck" buying at least one, good sized chunk of it.

This was not the only piece of land someone was anxious to sell us that had repeatedly failed perk tests and was NOT approved for building. Yes, people lie; even "professional" realtors don't necessarily convey all the facts or even know them.

LESSON #2: Make sure the land "perks" before you consider buying it. Check with the County Health Department (or the department within the local government with responsibility for approving septic systems) and they will tell you whether or not the land has been approved for a septic system and exactly where it has to be located and what type of system will be needed. Or if it's a piece of land "new" to the market, that has not yet been perked, the seller should pay to have a perk test done on the property which results in county approval. Don't buy a piece of property without official county approval of the perk test which will also tell you where you can put your septic field and what kind of system you will need. And, again, make sure the land perks for the number of bathrooms you have planned.

When we finally found our piece of land, and were told we could build a barn and a house, could have the number of horses we wanted, and that there was one area on the parcel of land that perked for the total number of bathrooms in the barn and house that we wanted, there was still something to learn. Because of the rectangular nature of our land, as well as the general lay of the land, we wanted to centrally place the barn and house. However, the approved septic field was at the north end of the property. As such, we had to build an **"engineered" septic system** rather than a gravity field septic system. That's all well and good, but the warning here is that an engineered field is significantly more expensive and the county agent is cranky about approving them because it requires more work on his part regarding inspections. As mentioned earlier, the engineered septic system requires the discharge from the septic storage tanks to be pumped out to where the approved septic drainage field is located.

I should also mention that our County Health Department when it does a "perk" test on a piece of land, is looking for a natural **seam of sand** located at some point between the natural surface (or top) of the land down to a depth of 20 feet below the surface. If no sand seam is found within that 20 foot boring of the soil, then there will be no septic approval. Also, if water is found (underground water including springs, rivers, etc.) during the test bore, then your land will not perk. You need to check with your county health department to get specific information on its perk test regulations.

What happened to us regarding the "perking" of the land should be noted as well in that it incurred a huge amount of money in order for us to actually have a septic drainage field.

First let me explain a little more clearly what happens during a perk test. In our case, the seller of our land hired an excavator to bore test holes on the property until he found a sand seam. The Health Department was present to witness the perk testing. A sand seam can be right under the top surface layer of soil or the sand seam can be at 3 feet or at 8 feet or at any number of feet below the surface soil—or maybe no sand seam at all. Remember that our health department required the sand seam to occur somewhere up to a depth of 20 feet at which point, if no sand seam was found, then boring stopped. Once the hole was bored, and the sand seam was located, water was then put into the hole, and the actual perk test was conducted to see if the sand seam was large enough to disperse the specified amount of water that would equate to the normal discharge from a certain sized home (based on the number of bathrooms). From that information, the Health Department issued a permit that specified where the drainage field had to be located, and established the number of bathrooms that would be allowed to use that drainage field.

Our county required that all the soil above the naturally occurring sand seam in our septic drainage field be dug out and removed from the hole where the field was located. The removed soil then had to be replaced with the approved type of sand required by the Health Department. Once the hole had been filled with sand, the septic field drainage pipes were laid on the field and then covered with a thin layer of topsoil that was added on the top surface in order to plant grass.

In our case, a sand seam wasn't found until 18 feet below the surface. That meant that we had to dig out an ENORMOUS amount of soil, have the soil redistributed on our land, and then buy an ENORMOUS amount of sand to fill the hole. The "heads up" point here is that process cost a LOT of money. Plus we had the additional cost of having to do an engineered field.

NOTE: So you might want to take into consideration your budget and decide now if your budget can handle a **perk that doesn't occur until a depth of 18 feet or more**. If you are on a tight budget, then it might be wise for you to include a further restriction on "land perking" contingencies in your purchase and sales agreement stating, for example, that the land must perk within 6 feet of the surface.

Also on our purchased piece of land, where in this township it was actually okay to have a barn with livestock in it and NOT have any human residence on the property (interesting), there was, however, another little surprise.

Shocker #2: Here it turned out, there was a **restriction on the number of horses** you could have depending on the amount of acreage. I had ten acres and four horses. The rule was that you could have only three horses on ten acres. However, because we were leasing the adjacent ten acres, we could have the four horses on our farm of ten acres. But the crazy part of the law was that after meeting the initial requirement of no more than three horses on the first ten acres, that if there was any additional acreage, be it one additional acre or one hundred additional acres, and whether we owned the additional land or we were leasing the additional land, that we could then have as many horses as we wanted on the other acreage! Just think about that for a minute. On one side of the fence we would have three horses standing on ten acres, and then potentially on one adjacent acre we could have a hundred horses or more. Go figure.

LESSON #3: Find out how many horses you can have on how many acres from your township or county office that controls zoning restrictions.

It's farmland, right? There were farms to the left of us, farms to the right. Yes, it was true that there were also a couple of singular, isolated, "neighborhood streets" scattered here and there among the farm fields, but overall, this was obviously a farm community and the land we were buying was obviously farm land. So when the realtor told us that "of course" we would be paying **farm taxes** because, anyway, that's what the woman who owned the land was paying (this phrase I now realize being the realtor's clever, little, point of deception), there was no apparent need for us to question that. Besides, why wouldn't farm land be taxed as farm land?

Shocker #3: Receiving our first tax bill which was huge and finding out we were zoned "residential in an agricultural zone" which meant we were being taxed as residential, even though we were on a farm in an agricultural zone, and being taxed on the **cost** (rather than market value) of all the buildings on our land including our house, barns and arena! You should be aware that there is little to no market value for agricultural buildings in conjunction with a residence, and therefore your tax assessment should reflect that reality, and if it doesn't, you should file an appeal.

Yes, we were on farm land. Yes, we were farming it--which meant in our case we were a farm because a minimum of 60% of our land was being "farmed" as pastures. Yes, the woman who sold us the land was also on farm land and was paying farm taxes. But what our realtor had failed to tell us was that the woman who partitioned off her land for sale had big dreams for her land and had planned to turn the acreage into an exclusive little neighborhood—that was until the land didn't "perk" except for the one, ten acre parcel we had purchased. (You remember that "perk" word, don't you?) Armed with her big plans, she had gone and had the parcels of **farm land rezoned to residential** in order to accommodate her subdivision plans which, although they subsequently fell through, left us being hit with maximum residential taxes on an actual, working farm!

Even more deceptive, was that the land was zoned as A3 where A meant agricultural (which made us think we were going to be paying farm taxes), but where the 3 that follows the "A" meant residential within the agricultural community—or in other words, a single residence house sitting on a lot of land. In this case A3 meant one house per 10 acres, whereas, for example, R4 meant 4 single residences per acre as would be found in town. Bottom line: A3 meant we were in farm country, but taxed as if we were living in the middle of town with all the services (fire, police, ambulance, road maintenance, etc.) around us. However, please note that paying taxes as if you are going to receive those "in town" services doesn't necessarily mean those services are going to be available to you outside of town. Obviously this little coding system is how the town or township gets maximum tax revenue, how the realtor gets away with saying your land is zoned agricultural, and how you get deceived.

To be more specific, in our case, we live only one mile from the center of the town where all the usual tax payer services are available, as well as the school system. However, our farm is actually "in" a different township that pays our township tax revenue to the town for the school system and all other services like police and fire protection. Although our township was taxing us as if we were receiving all the services from the "town," our "township" decided to save money and stop paying money to the "town" for such services as police protection—the township was now only paying for fire protection and the school system. As such, we were paying top tax dollars to the township, but we no longer received police protection. So, for example, even though we had a security system in our house, it was never clear who would be called if a break-in was detected. Can you imagine how surprised we were the first time we called the town police and they told us they didn't serve us and wouldn't come out—that we had no local police protection, but we could call the state police if we wanted. Have you ever called the state police and tried to get an immediate response? It doesn't happen.

Perhaps it would have seemed fair for us to be paying such ridiculously high taxes if the woman who sold us the land was also suffering for her little "trick" and having to pay high taxes on the other parcels of land she couldn't sell; but, no. She was not only still zoned as a "true farm" with low farm taxes, but she was further covered with a "**low income exemption**" which allowed her to pay no taxes at all.

NOTE: **Grandfather clauses** exist in some places to protect the rights of people who bought or built a house and/or farm under zoning restrictions, laws and ordinances that existed at the time of the purchase of the land and/or building or purchase of the structure, but may have changed over time. In general, as some basic restrictions, ordinances, and laws change, the current owner does not fall subject to the new changes (there are exceptions, however, like eminent domain). The ensuing laws, etc., that have been instituted since the old owner bought or built a property do not come into effect until a new owner purchases the property. This is a "**heads up**" for those of you thinking about **purchasing an existing property**. You should go to the township or county office and make sure there aren't any new laws and restrictions that have been passed since the original owner from whom you are purchasing the property built or operated the property in a certain way. You could find out that the property will be reassessed when sold and that the taxes will increase significantly. You could find out upon investigation that you no longer have the right to even have a single horse on the farm!!! You could even find out that the farm is now zoned "industrial!" There are other parts to grandfather's clauses that include such things as changing a private farm into a commercial farm. These laws, as an example, allow an original owner, after a certain specified period of time, to become a commercial facility without having to comply with current commercial structural requirements or handicap access laws. These laws should be made available to you through your township officials or the county. Confirming these laws with an attorney who assumes the liability should they prove wrong, would be important.

LESSON #4: Find out how the land is zoned and how it will be taxed when it is purchased by you, and what properties will be used as "equivalents." For houses, equivalents are houses of roughly the same square footage and quality of construction situated on roughly the same sized piece of land in a location that is similar and not too distant. Equivalents are hard to do for farms. In the case of empty land, the taxes will be based on how the land is zoned—not how the land "was" zoned while the current owner held deed to the property, but how it will be zoned once you purchase it. For example, our land was formerly farmed with crops such as corn and soybeans but was rezoned residential when split and sold to us and the taxes were significantly increased.

Then too, there's the question of accessing your land. Don't assume that you can just drive over anything including private roads and property owned by others to get to your land even if easements exist. Also don't assume that a little worn path across the land you've bought can be removed or even used by you.

<u>LESSON #5:</u> There are <u>easements</u> in existence and ones that will probably be needed for such things as electrical wiring for delivery of electricity. Find out from the County Hall of Filed Property Records and Deeds what easements exist on the piece of land you are considering purchasing. Specifically, three of the important things to establish are: (1) what easements would continue to exist if you purchase the property that would allow you to access your land (recorded easements can expire with language stating the date of expiration); (2) if you are allowed by zoning ordinances to use an easement at all to access your land or if, in fact, your driveway cannot use an easement for access to your land, but, rather, your driveway must directly abut a private or public road and not use an easement for access; and (3) what potential easements would continue to exist or might be imposed on your land once you purchase the land that would allow others to have easements ON your land. You need these and all information about easements, in writing, and all ordinances pertaining to the use of an easement for access to properties. Getting thorough information with protection of your rights and investment requires that you hire an attorney for due diligence and to assure that you are in full compliance with all ordinances.

Our seller--you remember, the one with big, subdivision dreams--had put in a private road which ended in a cul-de-sac to provide access to the parcels she was selling. That sounds convenient, doesn't it?

Well, it was and it wasn't.

The problem was that the private road and the cul-de-sac only served the first three, 10 acre parcels of land. The fourth parcel, which was the one we bought because it was the only parcel of the four parcels that perked, did NOT directly connect to the private road or to the cul-de-sac.

Further, her land (specifically the third, 10 acre parcel which was the ten acre parcel next to our land which was the fourth 10 acre parcel) totally blocked access of our ten acre parcel to the private road.

Recognizing that problem, when she made her land development plans, she took advantage of an old easement she used for the private road and that the original farmer used for access to the full 80 acre parcel she now owned that took 30 feet of the original farmer's land (now her land) that ran along the northernmost edge of parcel #3, and 30 feet of the southernmost edge of the adjoining neighbor's property to form a 60 foot wide access easement along the top edge of her property and the back edge of the adjoining neighbor's property. The existence of this 60 foot wide easement supposedly then allowed us to build a driveway anywhere within that 60 foot wide space in order to pass from the cul-de-sac over what was technically half "her" land and half the adjoining neighbor's land in order to access our land.

NOTE: I say "technically" owned the land in that the reality of easements is that even though you may own the land the easement is on, because others potentially have the right to use "your" land, the easement part of your land really becomes a "no man's" land. In other words, you "own" the land, but you really don't have any control over the use of the land, and without control over the use of the land, the question begs if you really "own" the land as such?

Although we would have preferred to put our driveway on the seller's 30 foot side of the easement to avoid any trouble with the neighbor who "technically" owned the other half of the easement, our preference was blocked by her driveway (that provided the only access to her house) being in the middle of the cul-de-sac adjacent to the southern boundary of the easement on the third, 10 acre parcel and the need to do a right angle turn on our driveway within the easement to enter our property. When blocked by the wide ditch on the adjoining neighbor's side of the easement, we were left with very little leeway as to where we could build an entrance to our property off the cul-de-sac that fit onto the easement. So based on where the entrance to our property had to be placed, and considering that we would have to widen and soften the right angle turn that occurred where the easement ended and then turned onto our actual property so that large semi tractor trailers such as moving trucks and horse trailers, and large service and construction vehicles could negotiate the turn from the easement onto our property, we needed to use some of the easement that involved the bordering neighbor's 30 feet of easement.

Ordinarily that would not represent any problem in that we were told that we had the legal right to place our driveway anywhere on the combined 60 feet of the easement from the two neighboring properties. However, just looking at the huge rocks and tree stumps placed along the rear property line of the adjoining neighbor, the 50 foot long by 8 foot high wall of interwoven tree branches and sticks on another part of his property line, and the barbed wire fence with old bleach bottles and orange colored plastic newspaper bags tied on it that completed the blockage onto the neighbor's property from the easement, there was a strong indication, right up front, that the adjoining neighbor wasn't exactly a welcoming sort of guy. We certainly would have preferred to avoid touching any part, of his part, of the easement.

An additional problem was that the seller's driveway, about halfway along our ten acres, took a shift to the left and actually cut into our property about a third of the way, and continued on our property all the way to the end of the ten acres.

Now we had a double problem. We couldn't get a wide enough driveway with a reasonable turn onto our property by solely using the seller's 30 feet of the easement, and the seller's driveway ran down the middle of our land for a significant distance making a large piece of our land essentially unavailable to us! And, of course, the seller refused to move her driveway.

Faced with these access problems, we "walked" away from the purchase when she refused to move her driveway.

NOTE: At the time we built our farm, the township here did not allow two adjacent property owners to share a driveway. They have since changed this ordinance. This would be another important ordinance for you to check out regarding shared driveways. However, shared driveways are another source of unending headaches and should be avoided if at all possible.

In time, the seller came back and she agreed to move her driveway off our land onto the adjacent 10 acre parcel. Unfortunately, she chose to run her driveway right between the two, ten acre parcels (parcel #3 and parcel #4) because we failed to specifically state in writing where she could put her driveway on the adjacent 10 acres. (Specify everything you can think of in writing regardless of how awkward you may feel about doing so at the moment, and regardless of what anyone says to you or promises you!)

I have to say that to this day it's kind of strange to be looking out over our peaceful twenty acre parcel (the owned land plus the leased land) to then suddenly have a FEDEX truck go zipping along her driveway right down the middle of the 20 acres! Then, too, there's my constant concern about our animals (cats, dogs, and horses) getting run over. Ugh.

Once the seller's driveway issue was resolved, we felt confident with the 60 foot wide easement that we would be able to put in our entrance to our property and our driveway with a nice, slow wide turn at the point where the driveway had to make a right angle turn onto our ten acre parcel. Even though we still felt uneasy about the adjoining neighbor's "Berlin Wall," we had been assured we could legally use any part of the 60 foot easement regardless of any personal issues the neighbor might have about our use of the easement. And so, we put in our entrance including a farm sign with our address, landscaping, mature spruce trees, three-rail white PVC fencing, and our driveway. We even first notified the adjoining neighbor, and actually got his "approval" of our plans which though unnecessary was done by us as a point of courtesy to him.

But, of course, this wasn't the end of the easement story.

After living on the property for more than a year, and having had the driveway in for more than three years, suddenly the adjoining neighbor with the Berlin Wall, on whose property half of the easement lies, came marching down our driveway to inform us that he "owned" our driveway and that he and his hunter buddies had the right to use our driveway anyway they saw fit including parking on it or next to it.

Oh, really? That was interesting.

He also informed us that because he had never signed a document allowing us to have a driveway on the easement, that our deed was, therefore, invalid.

Wow!

Well, if that was true—that our deed was invalid, and we had no right to use the easement--then our property would become valueless in that we would have no access to our property. Our farm would be landlocked.

This all got very complicated and created a litigation maelstrom for us to get quiet title. The problem arose because our attorney who advised us during the purchase of the land, and who drafted our sales agreement, did not properly do his due diligence and uncover some serious problems with the land. The point here is that you must be sure your attorney does his due diligence, reviewing all pertinent documents to get a clear, written understanding from the township or county zoning officials as to exactly what rights you have, and what the rules and regulations are on every part of your land or on any easement.

Some of the **"heads up" things to look for on an easement are**:

- Does the recorded language indicate that it expires or can be revoked?

- Does your title insurance have exceptions that exclude easements from coverage or use of easements from coverage even though they may be required for access?

- Who actually owns the land on which the easement lies?

- Who owns and controls any improvements to the easement be it a driveway, fencing, trees, landscaping? What right do you have to put and maintain these in the easement?

- Who has a right to use each and every part of the easement and its improvements, and how it can be used (foot traffic versus vehicle traffic as an example)?

- Who has the right to say "keep off" of what parts of the easement?

- Who will enforce these rights? Are these exclusive rights? Will litigation be required for enforcement? Who pays for the litigation? What circumstances enable title insurance to be collected?

- Who has the right to maintain the area and its improvements?

- If someone starts maintaining the easement even though the rights of use have been granted to you, does that person then have the right to claim the easement as exclusively theirs? (In other words, if your neighbor starts mowing inside the fence on the easement, does he then "own" that area above your right to exclusively use that area?).

Sometimes the local government will play games with you and refuse to put anything in writing and/or give you only partial information, i.e. only give minimal and exact answers to exactly what you asked and offer no additional, but most likely essential, information that should have been shared with you. That is mainly because small local governments are afraid of lawsuits being brought against them which they cannot afford to defend. Therefore, they are reluctant to put anything in writing or offer general information beyond exactly what is asked. However, they are the government and they MUST put ordinances, rules and regulations in writing. It's their raison d'etre, and so you must insist, and persist and get your attorney involved if necessary.

NOTE: This kind of a dispute over our right to access our land is an example of the type of challenges that are covered under the **Title Insurance** that you should pay for as an owner at closing and you must be sure that you have as an owner. Lenders' title insurance is different and you need owner's title insurance for the full value of the land plus improvements such as buildings. You must also make sure that you have a **warranted deed** so that in case it turns out that your title is not valid, you can then redeem the full price of your property from the title insurance company. Of course dealing with title insurance companies is even less pleasant than dealing with health care insurance, so don't be surprised at the attitude. In our case, we had an option to **purchase additional title insurance that covered the improved value of the property**— not just the land. I would recommend that you and your attorney read, and fully understand, and approve (in writing by the attorney) the title insurance policy so that you know exactly what is covered by the policy and what triggers a claim and payment before you purchase the property. In essence, the big idea is to transfer as much liability to your attorney as possible with careful written directions so that ultimately, if necessary, you can transfer liability to the attorney's malpractice insurance company should something have been overlooked by the attorney.

For the purpose of insight, I will include some of the items of interest we found in our easement documents.

First of all, we obviously own the driveway and have exclusive use of it. In our document there is a reference to the easement being used as a "public access." In our document that meant—singularly—that the police, fire, public utility trucks, trash removal service trucks, etc., that are needed to serve the landowners, can have access to the private road and the easement to serve us. That does not mean that John Q. Public can drive down the private road or the easement. It further means that an adjoining easement land owner cannot give rights of use to hunters or friends to use our exclusive private driveway or use the easement around our driveway for vehicle parking. This language for rights of use may have to be added to the recorded easement for clarification. However, there is other legal wording, that is not contained in our document, but could be contained in your document, that might mean that your easement is indeed available for use by the public—so you had better check out what your document means by "public access" and get that definition in writing.

Meanwhile, do not allow your neighbor to maintain your driveway or the land around it that you have "claimed" as yours via landscaping and fencing. If the neighbor "maintains" any part of the easement for 15 or 20 years, for example, (depending on your state laws), then those things you have put on the easement as improvements will become his property—that could be anything from shrubbery to fencing to sculpture. So YOU must maintain your "claimed" and improved part of the easement. If the neighbor will not cease and desist, you must have an attorney (or the township officials) write the neighbor a letter to stop maintaining the improved easement. In our case, the only reason a neighbor could use our driveway on the easement would be to access his property for purposes of maintenance on his "non-easement" land that adjoined the easement. That meant he could only drive down our driveway and then he had to immediately drive off our driveway to subsequently park only on his "non-easement" property, i.e., he could not park anywhere on the easement at any time. Obviously he couldn't damage our driveway or any improvements in the process. However, this language that clarifies rights of use should be in the recorded easement.

There are many other variations and types of easements including shared easements, public utility easements, and easements that can go away over time if you don't use them. Finding a lawyer who specializes in land/real estate can help determine and define any and all easements on land you are considering purchasing and such additional concerns as eminent domain rights of the state to do such things as potentially extend a road to go through a large portion of your property.

<u>NOTE:</u> By hiring an attorney to review all laws and local ordinances pertaining to your land before closing, and having the attorney "clear" you to build on the land by stating that you are in compliance with all laws including all local ordinances, is the best way to protect yourself if later problems arise. If your right to access your land or to have built on the land is later challenged, and you are found to have violated a law or an ordinance, then you can sue your closing attorney for malpractice to cover your losses. The "heads up" here is that there will be a statute of limitations on how long you have to sue your closing attorney which for us was only 2 years from the transaction but was extended to be 6 months after we first learned about the violation the attorney did not find before/at closing. You must file a law suit within that statute of limitations time period or get a tolling agreement (which legally extends the time period of the statute of limitations) before the statute of limitations runs out.

LESSON #6: Make sure that your rights of use of all easements are clearly defined in legal documents, drawn on, and stated in your filed, warranted deed to your property and that these documents including your owner's title insurance policy are approved by your attorney. These easements, as well as all the township or county deeds, will most likely be recorded in the county's recording office.

Then if your neighbor trespasses on the easement (where trespassing has been defined in the easement documents), you can call the police and file a report of the offense. Without a document that defines the use of the easement, the police will not come out and enforce a trespassing violation. The first step is to provide the police with a written clarification of who has what rights to use the easement, and what uses are violations as determined by an attorney, your title insurance company or other legal authority. Without this legal definition of rights, the police will not come out and enforce a trespassing complaint.

NOTE: Overall, I would strongly recommend NOT buying a piece of land that requires use of an easement or even a private road in order for you to access your property. Life is much simpler that way!

Next you should ask yourself some basic questions about the **location of your land**. If you're going to have animals, and especially horses that tend to be innately problematic and high maintenance, you have to consider your level of self sufficiency in the care of your horses. In other words, ask yourself if you can live on land that is in "the middle of nowhere." Yes, it's attractive to be way out in "the middle of nowhere" because it's quiet and peaceful and potentially because the land will be much cheaper. Nevertheless, the **self sufficiency** factor must be taken into consideration.

Beyond self sufficiency, now ask yourself what kind of access you will have to veterinarians, to farriers, and to hay and shavings suppliers if you will be living way out of town.

It's also important to remember that "crises" come in all sizes. Take for example something as simple as a horseshoe being "half" pulled off; your horse can't walk on it without damaging his foot. If you pull the shoe off incorrectly, you risk taking off so much hoof that your horse will be lame or unserviceable for a significant period of time while the hoof grows back in. So, you should honestly consider whether you can take care of even this kind of mini crisis without your farrier. Or perhaps you should consider that maybe you would be better off having your farm located where crisis services are readily available to you.

LESSON #7: If your farm is too far out of civilization, you may have trouble getting the best (or even any) veterinarian services and maybe not even a farrier.

And we're just talking about animal care and security. There are also "people" emergencies to take into consideration. Remember that "to ride" is to "fall off," and to "fall off" is not always to be able to "get up!" Ask yourself, if you fell, who would know that you had fallen? Who would take you to a hospital? How close would that hospital be?

These are the types of questions to ask yourself when you are considering where the land is located.

There is another important consideration when you're living "way out there," and even when you're living in close if you've got a lot of acreage separating you from neighboring land. You need to consider what the **wildlife** is in the area.

LESSON #8: Yes, there is wildlife.

Our land being in close to town, I never even considered for a minute that I'd see anything of note in the wildlife category that might threaten my horses or other pets. Nevertheless, the builder and the workmen kept "warning" me that I had better get myself some guns to ward off unwanted wildlife, and, specifically, coyotes.

Coyotes? I thought to myself. Well, that's just ridiculous. I'm not out in the middle of the wilderness. Why I'm just one mile from downtown.

But then there was that winter morning when I opened the slats of our bedroom window blind, and looked out across the pastures, and saw, walking along the end of the south pasture—right outside the fence - you guessed it … a coyote! Yipes.

Recently a man stopped by our farm who had "watched our farm go up" and wanted a tour of the facility. In the course of our conversation, he mentioned that he and his neighbors were having problems with coyotes over by his place (only about two miles away from our farm) and that the neighbors were organizing a "group hunt" to go after them. He said that his wife was shocked to have had a coyote follow her all the way up to the gate by the barn as she led her horse in from the pasture. He used that scenario to reinforce to us that coyotes "aren't afraid of us." He also mentioned that coyotes "bait" pets around barns and houses by sending in a single coyote to entice the pet to chase it, whereupon the single coyote leads the pet back to the pack where the pet has no chance of survival. Be warned. Be ever watchful.

NOTE: We have learned that in our state coyotes are considered vermin and as such there is open season on hunting them. That means we can kill a coyote on sight or go hunting to kill them regardless of the "official" hunting season. Also, in our state there is no hunting permit required to hunt coyotes.

And, of course, coyotes aren't the only wildlife. Depending on where you live, there could be several levels of critters to be permanently on the watch list.

But then too, what about the neighbor's dogs? Generally speaking, people who live on large tracts of land have big dogs with the intent of having the dogs serve to protect their property. As such, their big dogs, should they get loose, out of control or join up into a pack, can be a BIG problem that might also require you being armed with some kind of defense mechanism.

Now on the flipside of being out in the middle of nowhere are the issues of having your land closer into town. If you don't have a minimum of about one hundred acres to call your own, and insulate you from the rest of the world with wide distances for boundaries, then you will probably have the "joy" of **neighbors**.

I actually wanted to have neighbors from a security point of view if nothing else. I like to feel that if I'm in trouble there's at least a chance that someone might hear me calling for help or might come looking for me if I haven't been "seen around" for quite some time or note that there's something "not quite right" about the pattern on the farm--like a horse grazing peacefully outside the fencing, all tacked up with bridle and saddle, but without a rider! I also like to think that having random, human activity around the edges of the property helps keep down the presence of the larger wildlife that you would have if you were living farther out and were more isolated. I also happen to like people, and I like to wave hello and have an occasional chat. However, neighbors have brought unexpected and very upsetting problems that I never would have expected.

LESSON #9: Even "horse friendly" neighbors can wind up causing a lot of trouble, "innocent" and unintentional though it may be.

There are some people who won't want you and your horse farm as neighbors in the first place, and will be a constant source of trouble. They will try and blame you and your horses for any fly or rodent they see as if there had never been a fly or rodent before your farm existed. They will blame you for changing the lay of the land and causing water run off even though you can point to the gulch that has clearly been eroded from run off over the last, say 50 or so years! But that's just the hostile half of potential neighbors where you'll learn to just deal with each silly little issue as it crops up and put it to rest—mostly by ignoring it!

Then there's the other half of the neighbor types who are relatively horse farm friendly. Nevertheless, even this group carries some problems.

Probably the first and foremost consistent problem with the "friendly" neighbors is that they think you're going to let them come over any old time to "visit" you and check out the chances of you letting them ride your horses. That amazes me. But apparently thinking that anyone, with or without riding experience, can get on any horse and ride it and then walk away in one piece is only an "amazing" thought to people who are actually horse people and know better than to think that way. And, if you find yourself asking here what's wrong with letting the neighbors ride your horses, then I'm going to have to wonder about how far into horses you really are, and whether considering building a horse farm might be a little premature for you.

Note: Make sure your umbrella liability insurance policy covers injury to visitors from horses or other accidents on your farm.

But beyond the "bitterness" that the neighbors feel when they finally figure out that showing up with blue jeans, cowboy boots, a cowboy hat, and big smiles on their faces is not sufficient to convince you to let them ride your horses—particularly if you have hunter/jumper horses and English saddles -- here's a couple of other unanticipated "neighbor" problems that have arisen on our property.

I have two huge pastures plus one smaller pasture adjacent to one of the larger pastures. The barns and house sit in the middle of the fenced-in pastures such that we can refer to the two large areas as the "north" pastures, and the "south" pasture.

First came the problems in the south pasture. Yes, it's located a little farther from the barn—which always makes horses nervous anyway--and it requires walking past the house with its potentially spooky sounds and activities. Yes it's also bordered by woods on one side, thick brush on another side, and open land on the third side, and, as such, the pasture could give a vulnerable feeling to a grazing horse especially if they sensed wild animals in the woods--deer being a primary candidate in our case considering the large herd that frequents our area.

But our horses were used to seeing deer. They were used to pastures distant from the barn. They were used to houses and all kinds of people activity. They were used to grazing in pastures that bordered woods. Plus they had seen plenty of wildlife—and especially deer--over the years as well, and never over reacted. So, why was it then that now that the horses were on our farm, they were frequently breaking into gallops, they were clearly in a state of panic, and they were desperate to get back into the barn?

You tell me.

Well, my first misdirected inclination was to put the blame squarely on the chubba-lupalusly fat ground hog that was startled by me one morning when I raised the blinds in our bedroom window that overlooks the south pasture. I was certain that his little fat head perched in shock at my sudden presence upon his upright stance was THE thing that was taking the horses by surprise and making them run. Certainly this ground hog must be scuttling along on all four feet through the long, lush grasses of the pasture during its mindless "missions" to then erratically go from his four footed, ground hugging scamper into a sudden, and spook causing, upright posture to scan the pastureland. Seeing Mr. Ground Hog do that very behavior, I immediately fingered him as the source of my problem with the horses. I determined, right then and there, to find a way to discourage this evil invader from ever returning to frighten my peacefully grazing horses!

The first thing I did was to have my husband mow the pastures to the 6 inch minimum height recommended for pasture grass, while I mowed everything else as short as possible around the outside of the fencing. I also had my husband knock down and cut away all the underbrush and weeds that were growing around our pond, as well as those growing under the trees that bordered the woods' side of the pasture so that NOTHING could "sneak" up on my "poor" horses and scare the crap out of them—most especially, and in particular, Mr. Ground Hog!

Done! Problem solved ... well, at least to my mind and only for that moment in time. It wasn't more than two days later that I once again had panicked horses galloping across the south pasture and wanting to get out of the gate and back into the barn!

Okay, now I was at my wits end. Clearly it wasn't Mr. Groundhog who was the source of the problem because with all our mowing and clearing, there was no place left for him to sneak up on anything. But what was also clear, in the meantime, was that turning out the horses in the south pasture was over. The horses were a nervous wreck from this constant spook and run scenario, and so was I. I would just have to accept that there was some unknown, evil something lurking out near the south pasture, and that most likely that something was either the Loch Ness monster, living in our pond, or Sasquatch, living in the woods. But, either way, the bottom line was that at this point there was NO WAY those horses were about to go into that south pasture.

Months went by. The grass grew to a good two feet in the south pasture from lack of use (and lack of mowing). I kept vigilantly searching for that devil ground hog that I continued to believe was the source of all my problems. Then too, when I was feeling desperate for resolution, and I had not recently seen Mr. Ground Hog, and logic was NOT prevailing, I would scan the pond for Nessy, and then scan the woods for Big Foot. Nothing.

Then there was that late fall day when the leaves were all gone from the trees in the woods that bordered the south pasture, and I could now actually "see" into the woods, and I was behind the house in our yard playing fetch with our dogs. It was then when I looked up into the woods for a brief moment, that there in that moment, I could just catch a glimpse through the barren trees of our neighbor, who lived to the left of the rear of our house, walking through the woods with her two big dogs.

BINGO!

It WAS Sasquatch in the woods after all! No wonder the horses were panicking in the pastures. Remember, it wasn't until late fall that I could FINALLY see the neighbor with her dogs, because it wasn't until all the leaves were gone that I could actually see into the woods. Can you imagine how she and her dogs looked and sounded to the horses in the summer when the woods were dense and dark? Just to confirm that she (and not the poor little ground hog who, until that moment in time, had taken all the blame and condemnation) was the actual boogey man, the neighbor told me as we chatted that she and her dogs went for a walk in the woods every day!

Great.

But then, to my total chagrin, I started having the very same problem in the north pastures. I had already "lost" the south pasture to Big Foot and her dogs. Now I was getting Mexican Jumping Horses (a new, designer breed of horse, I guess you could say at this point) in the other pastures. What in the world was going on now???

Here was another HUGE pasture that should have had the horses grazing way off and away from the barn, but had the horses all huddled together and panicked at the gate. And on this northern side, there was no wooded area where the Big Foot pack could secretly roam nor was there a pond for Nessy to stealthfully navigate. No, in fact, this was relatively open land that bordered the big pasture on the north end. So what in the world could be wrong now?

Well, on one of those "special," "horses-all-huddled-at-the-gate" days, I decided to walk up the property line to the end of the pasture to try and see what in the world the source of the problem might be.

When I reached the top of the pasture, I couldn't believe my eyes. What did I find? Nothing less than my neighbor digging in a deep hole in the ground practically the depth and width of an Olympic sized swimming pool (without water of course), AND wearing a frickin' bright orange baseball cap!

"Oh, hi," he casually greeted me.

"Hi," I returned. "What are you doing?" I managed to ask as politely as possible through what was rather clearly my total shock and disbelief.

"Oh, this is my hole," he answered.

I just blinked and stood still waiting for more information before I reacted.

"This is how I get my exercise," he eventually clarified. "You know, some people go walk on treadmills at health clubs, but I come out here and dig in my hole."

Oh, Dear Lord. My neighbor, who lives on an equivalent sized plot to my twenty acres, with only part of the back edge of his twenty acres running along the top of my north pasture, had decided that of all the places he could dig a hole on his twenty acres, that he should dig a hole right there next to my pasture.

And quite a hole it is! In fact, it's a hole deep enough for his entire body from the neck down to fit in which then, of course, leaves "just his head" visible to bop along the top edge of the hole as he digs and digs. As such, that means when the horses look out across the pasture from the entrance gate, what they see, when the neighbor is digging in his hole, is the neighbor's head—only his head—just his head bobbing along the top edge of his hole. In other words, all the horses see is, what is to their mind, a walking head—a bodyless, bobblehead, if you will.

Horses have enough trouble dealing with fearing just about everything. Can you imagine how afraid "Mr. Bobblehead" was making them considering that they had no way to categorize a bobbing head in their list of potential predators versus benign hole digging neighbors—not to forget that the bobbing head was donned with a bright orange cap! Frankly, if I were a horse, I think I too would categorize this neighbor as a potential predator!

When I politely suggested to the neighbor that his hole digging activity was SCARING THE HORSES TO DEATH, the neighbor curtly replied, without any eye contact, that "they'll get used to it" which of course was meant as "passive aggressive code" for "go to Hell!"

Okay.

Well, in the meantime, the horses haven't gotten "used to it." It's been more than a year, and I'm still having trouble with Mr. Bobblehead and the horses. One visitor to the farm (who witnessed Mr. Bobblehead in his hole and the horses "freaking out" even though they were in the barn in their stalls!), suggested planting elephant grass as a visual block which he had seen grow thick and tall in South Vietnam. Unfortunately a little online research turned up the fact that elephant grass only grows in a tropical climate which is not what I have here. So I decided to go to a local landscaper for suggestions as to what to plant who, in turn, after discussing the situation, suggested we "just put up a solid, wood, privacy fence." Another potential solution would be to install an earthen mound along the property line which I have seen many people do.

What have I done about the bobblehead situation? Well, to date, I just try and avoid the times when I "think" Mr. Bobblehead will be out there in his hole, and just keep the horses in the barn.

But then there's **hunting season**.

I knew the neighbor behind us was into guns. He made a point of shooting them during a target practice session the very first morning we moved onto the farm, right after I had turned out the horses. I also knew he had built a deer blind out of fallen trees and brush in the woods right behind our house which is not only dangerous, but illegal because it is too close to our house according to hunting regulations which in our state say that a safety zone exists within 450 feet from any building or house used in a farm operation. But what I couldn't figure out, now that it was hunting season, was where the frickin' guy was in the woods and once again scaring the poop out of the horses. That was until the day when I finally saw him hugging the bottom of a tree in the woods about 20 feet from the south pasture and taking a last step down from the tree to the ground. It was then that I also realized that "what was now down, must have just been up" somewhere. So, following my logic flow, I proceeded to look up the trunk of that tree Mr. Hunter was hugging, whereupon I finally saw "it."

About twenty feet up in the air was the "source" of the trouble—a "hunter's" seat he had nailed into the trunk of the tree, way up "there." Okay—it even spooked me when I finally saw it. Can you imagine how spooked the horses were to see a man sitting up in a tree? Or coming down from it? Or climbing up to it? Yipes.

Even the friendly gesture of neighbors giving your horses apples can turn sour. Mr. Bobblehead also has apple trees. He took great pleasure in merrily walking down our long driveway passing out apples to the horses as he went by them on his "daily walk"—or what I eventually came to realize was in reality a daily "snooping expedition." But, "how nice," I thought, "to give the horses apples."

Then one evening when I went to close down the barn, I noticed that my thoroughbred was definitely not feeling right. He was standing motionless in the back of his stall, his eyes were half closed, and he had no interest in moving. As soon as I saw his distinctive demeanor, my head screamed "colic!" I immediately entered the thoroughbred's stall and tried to get the horse to interact with me and move. He wouldn't. Then I checked over his legs and body to make sure there wasn't some swelling or heat or other obvious injury causing the problem. Nope. He was definitely in a state of pre-colic.

But now my mind raced to think of why this horse would suddenly be in such a state. What could have happened? What could he have eaten? I hadn't given him anything new to eat. He hadn't been anywhere new that he could have eaten something strange. So what in the world was wrong?

"Apples!" my head suddenly screamed back at me. "Oh, dear Lord, that's right, the apples! It must have been the apples! That's the only thing that's been different!"

Yes, indeed, I now remembered that the neighbor had in fact come into the barn earlier that day and fed apples to all the horses—lots of apples—and had even left behind a bucket of extras. I quickly went over to the bucket of apples and sliced one, and then several others, in half. And what did I consistently find after slicing several apples in half? Worms. Lots and lots of worms.

Need I remind anyone reading this that horses are herbivores? Worms are definitely not on their menus.

I ran to the tack room to get Banamine, then gave my horse a dose, and walked him up and down and all around - until, in time, he felt better.

NOTE: Even if your vet is relatively close by, you should ask him to give you a supply of **basic medications** to have on hand including bute (phenylbutazone) for pain and Banamine (flunixin meglumine) for colic. (I prefer to use the oral paste form of both of these medications.) You should also have a horse "emergency" medical kit with basic supplies in it such as Furacin spray, a medicated ointment, an anti thrush product, iodine, isopropyl alcohol, alcohol wipes, vet tape, sterile, non-stick, gauze, etc.

Even though I had saved my horse from full blown colic in this apple eating instance, I was still left with the unpopular need to say something to the neighbor about the apples. Can you imagine how well that went over? Not good. And amazingly, that wasn't the last of the neighbor's tainted gifts to the horses. Months later he left a bag of carrots in the barn office with a note that he had so many carrots that he and his family hadn't been able to eat them all. My mind immediately made a link to the probability that there was an unstated time factor regarding the freshness of those carrots in that note. I opened the bag - sure enough - the carrots were past date, smelly, and very slimy.

But, I digress. Let's move on now. After all, neighbors do have their positives too--although I have to say that I'm still searching for ones big enough to balance the negatives that have presented themselves to date.

The next to last thing to consider about your piece of **land is where it "sits"**. Is your land sitting right alongside a main road? Or is your land sitting a short distance back from the road where you will have more privacy? Or is your land so far back from the road that no one can really see you? There are positives and negatives to each of the situations and which piece of land you purchase will depend on how you feel about each of them. To help you make your decision, I will play devil's advocate again to start you thinking about what you really want.

Remember that most people immediately assume, when they see horses and/or a nicely kept property with horses, that you are rich. Although most people are good and honest, there are also plenty of opportunists in the world, as well as criminals. If you are sitting "pretty" right along the road, you, your house, and your horses are highly visible to everyone. Plus, somehow, people feel they have some kind of a right to "invade" your privacy when you sit right along the roadway. You also won't have much privacy in that you are essentially right in the face of everyone who drives by your farm. However, the upside is that you will only need a short driveway on your property, you can easily access the main road regardless of the weather, and you can be easily seen by passersby if you need help.

If you are back off the road far enough that you can still be seen, but not so easily accessed, you are in a good situation. You will be able to have substantial privacy, but be close enough to the main road so that you aren't isolated, and you can be "seen" if, and when, you need or want to be seen.

If you get so far back off the road that no one can actually see you, then you will have total privacy, but you will not have the security of the public being able to see you should something go wrong. You will also be more open to robbery or other exploitation in that there isn't anyone driving by whom might note criminal activity occurring. Additionally, long access roads needed to access your property could be very expensive and hard to maintain and potentially inaccessible in bad weather. Our house sits back about 1500 feet from the public road.

Okay, so now you have a piece of land that seems to have checked out pretty well for your intended usage, the taxes would be appropriate, it seems to be properly priced, and you can afford it. By the way, the **pricing of the land** is an upfront clue as to how you are going to be taxed. Remember that you can be "residential" within the "agricultural" zoning. That's how we got snookered about taxes. The realtor wasn't exactly being "dishonest" when he said we were zoned agricultural. He just failed to mention the rest of the information about being an A3 rating which means we were considered residential within an agricultural area.

Farm acreage should be about one tenth to one fifth the cost of residential acreage. Use that as your general guide. In other words, if you are paying a lot of money for the land, then you are probably on land that is zoned residential in one way or another and you will be paying a lot of money in taxes.

Your last step should be to **go out to the land and "envision"** where you want to place such things as:

- house, garage and surrounding yards
- horse barn
- indoor riding arena (and whether it is connected to the barn)
- hay barn
- muck rack or other manure disposal areas
- trash containers such as dumpster with truck access
- other "out" buildings (for example, for equipment storage)
- pastures, paddocks and round pens
- fencing and landscaping with trees
- outdoor riding rings
- outdoor wash racks
- all gravel or asphalt roads within the property and access driveways
- hay or grass fields if any (that can be used for parking vehicles)
- gravel access pathways to pastures
- septic system tanks and field (they may be separated)
- well and water storage tank (if needed)
- propane storage tank if needed (and whether it will be buried)
- parking lots for things like cars and horse trailers
- ponds that may have to be created
- water drainage flow areas
- pathways for utilities - electrical and telecommunications

This would be the first step in creating a **site or plot plan** .

Be sure to also consider the direction of the prevailing winds across your land, as well as where the sun rises and sets, which will be important for your horses both in and out of the barn as the seasons change.

Orienting a barn with a center aisle and stalls on both sides of the aisle so that the barn aisle runs north to south allows the stalls to then face either east or west depending on which side of the aisle they are located on. Having the stalls face either east or west allows in a nice even distribution of light into the stalls and the barn during the day with the eastern side stalls getting the morning sun and the western side stalls getting the setting sun. In such a center aisle barn, you would not want to have a north side where it is sunless all day and a south side where there is sun all day—especially in the summer when the heat from the sun could become overwhelming.

However, the orientation for sunlight will depend on the configuration of your barn. For example, if you have just one row of stalls off an aisle in your barn, then you might want all the stalls to face the south and give light to the stalls throughout the entire day, especially in a cold, northern climate. The point is to be aware of the barn's orientation when you are envisioning the placement of your barn in order to meet the wind and sunlight needs you feel will best serve your horses.

Once you've "envisioned" everything at the site, go home (or do it onsite if you were clever enough to actually remember to bring along a pencil and a pad of paper!) and **do a rough sketch of a site plan** that describes where you want everything as the basic starting point of all planning and construction that will follow. You will need to have a feel for the perimeter measurements of all the structures in order to be able to "place" them in even a rough form on your drawing, and then to eventually be able to stake the structures' placement on your land. Later in the book I will discuss suggested sizes for barns and arenas. I am assuming here, however, that you have some idea of what style and size of barn (and house if you are building both) you want at this point even though things may radically change as quotes come in, as they did for us. However, the change in size usually involves downsizing (which I will assume here), and a site can usually accommodate a smaller barn or house than that originally planned.

I should take a minute here to note that a lot of people start out their farm really small with just a **run-in shed** and one little paddock (small pasture) encircled with wire or tape horse fencing. This is a common approach to an eventual "horse farm" and does have some positives associated with it—and of course a bunch of negatives.

The positives would be that doing a quick and small, minimal facility would allow you to:

- move your horses onto your land right away;
- have more time to find satisfactory builders;
- get a feel for the land and where you should do the final construction of buildings;
- see where the rain water and snow melt water run off occurs, and determine whether or not you need a pond;
- get a feel for the neighbors (if you have them) and all the other "wildlife" in the area as well!

The negatives include such things as:

- your horse/s having minimal protection from predators and weather;
- the overgrazing of the paddock which will eventually kill off all grazing grass and create the mud factor;
- no storage area for hay, grain, tack, tools, etc.;
- no dedicated space to conveniently work on or care for your horses.

Photo #2: Our Land at First with Just the Construction Driveway

However, don't forget about the zoning laws which may completely prohibit building an animal shelter and/or having animals on your land without your primary residence first being built on the land.

But let's get back to planning and building a complete barn and arena.

Although I don't know your land or your ideal farm plan, I can offer you my experiences here on what was involved in **my planning both for the land and the placement of the buildings and the pastures** on our farm. I offer this information as a "brain jolt" to get you thinking about your own land usage and building placement planning.

Photo #3: Overall Flatness of Our Land

When my husband first drove me onto the piece of land that is now our beautiful farm, I was not overly excited. But we have built a lot of houses over the years and there are some basics that we look for in land that this parcel had.

The first things that struck me about our land were the abysmally boring, overall flatness of the majority of the property, the thickness of the weeds covering the land, the clear low spot where water accumulates at least periodically, and a perimeter growth of mature trees, a lot of which were dead, killed by the ash borer; an unwanted "gift" from international trade.

Photo #4: Big Tree Spade Truck

Not a pretty picture.

However, not all that bad either.

First of all we have learned over the years that it's better to have a piece of property without any **trees** than it is to have one covered with trees. Why? Trees get quickly in the way of your buildings and have to be removed which is expensive and difficult. Trees also block the sun so that grass, at best, doesn't easily grow under them, and maybe won't grow at all. And then too, trees need constant trimming and occasional felling.

Yet, trees are beautiful. They both keep you in touch with the changes in daily weather and the changes in the seasons. They provide shaded areas to cool you and your horses in the summer and block the winds in the winter. Plus, they are good for planet earth!

So what is the solution? It's simple—again, not cheap—but quite simple. You buy "big trees" from a "**big tree nurseryman**" and put them where you want them on your cleared land. "Big tree" nurseries are NOT the kind of nurseries you see as you drive along the roadways with cutesy, little garden "knick knack" filled stores in the front, surrounded by thousands of potted plants, and backed by a few rows of balled and bagged, one to three inch trunk, trees. No, these "big tree" nurseries have to be found in the yellow pages. They are "tree growers" and they have acres and acres of fairly mature, medium size trees growing in fields far off the beaten track. These are people you have to call

Photo #5: Big Tree Planting Norway Maple

Photo #6: Norway Maple Tree in Pasture

and make an appointment with to meet in some obscure location where you had better be prepared to drive considerable distances between different fields for different types of trees, and then to get out of your car and walk long distances, at times through mud and weeds, to find your trees. But the advantage is that you get to buy beautiful specimen trees of considerable size (up to about 6 inch trunks for deciduous trees and up to about 20 feet in height for fir trees), and then have them planted on your property exactly where you want them after all the buildings and driveways have gone in. These "big trees" are planted with a special, huge, truck-mounted spade. The spade truck will first come to your property and dig the hole in which your big tree will be planted, take away the dirt from the hole, and then return with your tree which will be fitted perfectly into the hole.

We'll get back to tree selection later.

So it was actually good that our property didn't have trees on it. What was bad was that there was a clear **low spot** that showed all the signs of major water run off across the property and the trees contained therein showed signs of periods of deep, standing water. What was good about that, however, was that there was a clear low spot on the property which made it easy to determine where we needed to dig a pond. What needed to be determined, however, was the source of all the run off water and how much there might be in a "bad" year. We would need to be able to channel all the run off into the pond we would dig and the **pond** would have to be big enough to accommodate all the run off water even in a "bad" year. That is not so easy to completely figure out.

NOTE: Please resist the urge to build your barn (or house) at the low point on your property to be near the picturesque little **stream** that flows through the land in order to have that little stream run by the barn. Why? Because there will come that one spring or summer day when all the natural elements combine together and you will find both you and your horses standing "in" the stream, rather than next to it! Nor do you want to build your barn at the bottom of flowing hills and meadows where snow melt and rain water WILL run off and accumulate and make temporary, but fully flowing rivers, and standing water lakes, from time to time!

Photo #7: Drainage Pond -- First Dug

A good place to start to find out about the reality of **run off water** across your land is to ask the long term neighbors what they know about how the water flows during and after a heavy rainfall, during the spring thaw, during a "wet" year, and anything else they might have happened to have noticed. Sometimes they know a lot.

Remember Mr. Bobblehead? That hole he digs in actually does have a purpose—it's his run-off basin. However, he has no real reason to have one located where it is in that the actual low spot is on our property and is the place where an additional pond should exist. Nevertheless, his hole still does take the bulk of the run off water in that area and as such he saved me from having the expense of putting in two ponds.

Our ten acre parcel of land is slightly higher on the north end with a gentle slope (other than the one corner where Mr. Bobblehead has his pond) of the land to the middle of the land where it is flat for a considerable distance. After the midpoint of flat land, there is another gentle degree of slope to the end of the south pasture. However, in the one section of the south pasture, over by the woods, there is a more severe degree of slope and it is where the low spot naturally occurs and is therefore where we dug our pond. Considering the natural lay of the land, we decided to situate our barn and arena right in the center of the property. We decided to put our house over near (but not "next to") the low spot so that we could have a walk out basement in our house. (It did however make me nervous—and still does—that someday there might be "water problems" because we are near that low spot.) The house and barn are only 150 feet apart. Then we placed our hay barn 124 feet to the left of the main barn end door (67 feet from the arena end door). The pastures were split between the north end of the

Photo #8: Drainage Pond -- Mature

property and the south end of the property: specifically, there were two north pastures (one large and one small), and then one huge south pasture.

NOTE: Remember that it is illegal in most places to change the **lay of the land** that causes a change in the natural run off of any water onto your neighbors' land. You can accumulate run-off water in a pond or catch basin, but you cannot redirect the natural flow or block the natural flow of run-off water.

Having a centrally placed barn, placed close to the house, has worked out perfectly. No matter what the weather, it's **a quick walk or run over to the barn**. I can also see into the barn aisle from several of the windows in the house which is reassuring. Of course the horses in turn can also see me and therefore let me know they're thinking of me each and every time they catch a glimpse of me passing by a window by letting out a loud whinny that is clearly meant to demand both my attention and more hay!

Now take another look at your land and think out about your views, your building placements, and your pastures.

If your land is ironing board flat, ask yourself if that is what you want to see each morning when you open your front door and walk from the house to the barn. Or would such flat land eventually become kind of boring—maybe both for you and for your horses? Or perhaps your land rolls so much that your horses can't get a good, long, flat run anywhere on the property to release their extra energy. Is that what you want for your horses? How about where your house will sit? How about where your barn will sit? Are those pieces of land flat enough to accommodate the building without much need for bulldozing or additional soil? Are they high enough that water won't run back into them? And, again, look to see if water is ponding on the property or if there are any signs that it has ponded, or signs of established water run-off areas (start at an obvious low point and look out and up for long, eroded, pathways etched into the land).

What about **elevations**? Think about all the relative elevations where you have placed your buildings et al? Is your house sitting on a piece of land at an equivalent height to the height of your barn's site? Or is your house sitting lower than your barn such that potentially the second story of your house is at the same level of your one story barn thereby making it look like your house sank into the ground or as if you have to enter your house through a second story window?

Do you want your house to sit lower than your barn so that your barn dominates the landscape? Or do you want the house to sit higher than the barn so that your house dominates the landscape? Or do you want the house and the barn to sit at the same height on the land? This is the same kind of thinking that you must consider for all the buildings on your land. And then, what about your driveway? Is your driveway running downhill into your house? If so, that could be really bad news in that you most probably have created a chute for rain water and snow melt to flow directly into your basement.

Ideally, all your buildings should be sitting high and dry and at relatively equal height across your property, even if there is rolling land in between these high, and level, building sites.

NOTE: You will most likely have to move at least some soil around on your property to accommodate all your sites. But, other than that **soil distribution**, please remember that you can also choose to move dirt around or add dirt to make a flat piece of land have some mounds which will add subsequent interest to an otherwise flat piece of land. You can also flatten out land that has too many natural berms to make it more horse friendly. Once again it is "just" a matter of money to make your land look anyway you might want it to look. Remember that generally speaking what you cannot legally do is change the natural flow of water across your land or concentrate it in a way that affects the flow of water across neighboring properties

Now let's talk a little **more about pastures**.

One point to consider when envisioning your pastures is that there are both advantages and disadvantages to having totally separated pastures, meaning that half the pastures are on one side of the barn/house area and half are on the other side. On the one hand, the separate pastures allow you to correct for problems you're having on one side of the farm by being able to move your horses to an entirely different pasture experience on the other side of the farm. Like when Mr. Bobblehead is digging in his hole outside the north pasture, I can still turn out in the south pasture. What is bad is that a single horse turned out in the south pasture can get lonely and nervous. Also, it is difficult to keep an eye on the horses in the pastures when they are located at either end of the barn because it requires a walk down the aisle first in one direction, and then another walk down the aisle in the other direction. And rest assured, that there will be days when by the time you leave the pasture check on the one end of the barn, where everything was all calm and peaceful, that by the time you get back from checking the pastures on the other end of the barn, that if you look out again at the first pastures, all hell will have broken loose in the meantime!

Also, as I mentioned, a single horse in a single pasture gets nervous. Therefore, it's better to have at least two pastures next to each other, but not sharing a **common fence line**, so that your horses have company at least in an adjoining pasture. You should have at least 25 feet of **spacing** between any pastures to prevent "over the fence" fighting which, although more expensive in fencing costs, will cost you less in injury to your horses. I would also suggest having a minimum spacing of 30 feet between your **property line** and any pasture fence on your property, as well as between any pasture fence and any densely wooded area or roadway, in order to give the horses a nice safe feeling of space between the pasture and any distraction or surprise a neighbor or traffic might cause. Having a good, 30 foot spacing will also give you plenty of area in which to install a visual blocking scheme should the neighbors prevail in disturbing your horses. Also useful is having varied pasture sizes. Smaller pastures are good for sacrificial ones in the spring muddy season if you need one. Bigger pastures are great for playing and running as well as for turning out in large herds. However, if the pasture is too big, and especially if you have hunter/jumpers ...

Well, let me tell you how it went the first time. I had turned Emma, our senior Irish Hunter mare, out into the big south pasture and knowing how much she LOVED to graze, and how she was ALWAYS turned out alone, being the grumpy mare that she is, I thought for sure that she would be in horse heaven. I merrily walked back to the barn feeling good about myself knowing how happy I had surely made Emma by putting her into that huge, lush pasture. Leaving that happy thought behind, I returned to the barn and delved right into my newfound "joy" of "professional stall mucking." I was feeling all special about everything even though the outside temperature was about 95 degrees with a 90% humidity factor, and as I mucked I was sweating like a fat man in a wool suit. I worked along a good 20 minutes at my task, when in the back of my happily self absorbed mind I thought I heard a "funny" noise. It was one of those kinds of noises that didn't instill instant panic for action in you, but a noise that just kind of keeps nagging at you that maybe something isn't quite right. But I resisted the temptation to investigate the source of the noise in that I was almost finished mucking out Emma's stall and I wanted to finish the job.

But in those final few minutes I had grabbed onto in order to finish the stall, it was then when I heard the unmistakable sound of metal hooves hitting hard and fast along concrete. Recognizing the sound, and the potential problem associated with it, I immediately turned around and looked toward the aisle and saw ... you guessed it ... Emma flashing by me, right down the middle of the barn aisle, in, shall we say, a "slightly" modified gallop.

I couldn't believe my eyes. I thought to myself, what in the world was going on here? Had I forgotten to close the pasture gate? How in the world had this horse gotten loose? And why was she running? Could I have been so air headed that I left the gate open to her pasture?

Apparently I thought that was the strongest possibility, because rather than go try and locate and catch Emma, I first left the stall I had been mucking, walked down the aisle and looked out to the south pasture to see if I had indeed forgotten to close the gate. I was stunned to find that the gate was firmly shut. I thought, "What in the world have I done? Did I turn out the horse on the wrong side of the gate?" Yup, I actually was that confounded to try and figure out what had happened that I asked myself that ridiculous question!

Then my eyes as they tried frantically to help my stupefied brain regain its function, fixated on the drooping hot wire along the top rail of the fencing in the pasture Emma had exited. That's when my brain finally caught up with all the obvious clues my eyes were spreading out in front of me.

Yes, indeed, my 19 year old horse, the gray mare, the mare whom we had stopped asking to jump six years ago because we thought she couldn't do it anymore, had actually jumped out of the pasture—out of the four and one half foot high pasture fencing. The realization of that event was amazing enough to ponder. But what my mind immediately begged an answer to was the next most obvious question, "Why had Emma jumped out of the pasture?" Well, finding the answer to that question took me many months to figure out as I have previously mentioned. You do remember Sasquatch, Nessy, and friends, don't you?

Photo #9: Emma, the "Old Gray Mare" -- Maybe

Yes, most horses will, and are capable of, jumping out of any pasture. However, giving horses a pasture with a really long run does facilitate their natural athletic ability to do so, and almost seems to encourage them to take that option when frightened which in turn can subsequently result in frequent fence jumping. This is especially true when you have the kind of inspirationally, scary neighbors and wildlife bordering your property as I do.

Oh, and as far as my overall comment regarding Emma, I guess I'd have to say that the old gray mare "IS" still, unquestionably, what she used to be!

Again I would recommend that you put your pasture fence lines well in from your property lines to minimize neighbor issues and provide a wide maintenance corridor.

And, if you plan on turning out horses in herds, then you will want to plan in at least two, very large pastures. However, if your horses are turned out singularly, then having smaller pastures, with at least two pastures near each other, is a good idea.

Regarding "envisioning," it could happen that you are just not capable of envisioning such things as pastures and building placement which is not unusual. In fact, that's exactly why equestrian farm planning consultants, architects, site developers, and civil engineers exist. However, needing to use any of these services means having to spend more money, and it could be a significant amount of money. But, rest assured that there are plenty of people who would be happy to have you hire them to help you with such envisioning.

A quick example follows for those of you who have money to burn and might recklessly hook up with exploitive people who would like to ignite your cash for you.

While touring a horse farm/house complex just built for a multimillionaire at a cost of $16 million, and for which one of our acquaintances had done some of the interior decorating, we were introduced to the builder. Upon hearing that we were considering building a horse farm/house complex, the builder's ears perked up and he immediately offered his services. We in turn immediately informed him that we would be working on a budget and that it wouldn't be anywhere near the budget that had built the horse farm/house complex we were currently standing on. He in turn assured us that he had built homes on all levels and that he really wanted to discuss the potential of building our farm. So with the continuing pressure in our faces from the builder, we ran a square footage number past him right then and there for just the house we were planning and asked him to offer his cost for that size house to give us a feel of where he was really headed. He did a quick calculation in his head and came back with a ridiculously high-end estimate which immediately confirmed to us that he wasn't the builder for us. No surprise. We then tried to slough him off and go mingle with the crowd. However, he was aggressive and we couldn't get rid of him nor could we understand why he wouldn't go away after we had told him he was outside our budget.

A couple of weeks later, the builder called my husband and again said he wanted to do a bid on building our house. My husband again tried to slough him off, but once again the builder was persistent like a Jack Russell terrier clamped onto a mailman's pant leg. He pressured and pressured until he finally made an in road to my husband by saying that he could give us a better feel for a project with him if he could just come out and discuss the land with us on the actual piece of land. And then, when my husband seemed to be willing to take that first, seemingly benign step … the builder asked us to please have a check ready in that it would cost us $10,000 for him to come stand on our land and recommend where we should place our house—and only the house—no mention of barns.

Now if you have that kind of money to spend, and can't figure out what you're looking at land wise or you can't envision what you want to build and where, then maybe spending $10K on such a builder works for you. But it didn't work for us. For those of you on a budget, like us, the point here is to be careful of whom you are dealing with and potential undisclosed costs.

Well, moving along, now that you have a rough plan of where your house, barn, arena, hay barn, other structures, and pastures will be situated, it's time to figure out how you are going to access all those buildings. It's time to think about **driveways and pathways**—and do please note that both those nouns have an "s" on them.

Photo #10: Aerial Photo of Farm Showing Driveways and Pathways and Angled Corners of Pasture Fencing and Full Perimeter Fencing

A farm is about more than having a driveway that enables you to cross your land to drive up to your house. **Building a farm is actually like building your own small town.** As such, you must plan a complete road system that will allow access to all your buildings and pastures and service areas for all types and sizes of traffic from vehicular to foot (or hoof!) traffic. (As an example, eighteen wheelers have an average length of 70 to 80 feet and need a road width of at least 27 feet to do a 90 degree turn.) In addition to major driveways, you will need to envision every place you think you might need an access pathway—like the walk to each pasture, the entrance into the ends of your arena, parking areas for visitors/boarders to the farm, horse trailer parking pads, and heavy duty truck turnarounds.

Right now you don't have to worry too much about elevations in regard to your road system, unless there are significant low or high spots which you should avoid anyway. The low spots probably will require culverts. Whereas minor variations between original and final grades can be corrected without too much additional trouble and expense—well, relatively speaking - always remember that moving or adding large amounts of dirt will require large amounts of money.

In fact, now that we're talking about it, a **construction driveway** is one of the first things that actually has to happen on your land. Why? Well, obviously without one, there is no actual access to your land. Maybe you're driving an all terrain, four wheel drive vehicle, but workmen's trucks, semi tractor trailer trucks and cement mixers don't have those features and will need a driveway. It is important for you to understand that driveways are built in phases: first a construction driveway, then repair of the construction driveway, and last the finish gravel over the construction driveway—plus an additional top layer of asphalt if you choose to asphalt your driveway which I don't recommend and will discuss later.

One of the more brilliant things that my husband came up with on our farm, was a **turnaround loop** at the north end of the barn, and in front of the hay barn, plus a widened driveway for a parking area in front of the horse barn. These two areas have allowed all the large semis and service trucks ample room to enter the property, turn around in various versions of K-turns near our horse barn, and then leave the property. If semis can't access your property, you will find it more than inconvenient when the truck driver "drop ships" your fencing material order, for example, or other heavy delivery, at the entrance to your very long driveway, and it becomes your responsibility to get all that heavy stuff down to your house and barn. You will also be unable to have manure and general trash removal near your barn or house if a large garbage truck cannot come down your driveway to service a dumpster.

LESSON #10: If you are feeling pretty serious about purchasing a piece of land, and the preceding basics have been covered, including a fully sketched out site plan, then now is a good time to have the property surveyed in two ways: establishing the property lines, and topologically.

Let the seller know you are interested in the land and that you want to have it surveyed. **Surveyors** are reasonably priced. You will have to survey the land and have the boundary lines marked before you can purchase the land anyway. The cost should be no more than a couple of hundred dollars. Just having the boundaries marked on your land may be sufficient for you to determine if you like the land, where you want to locate your buildings, pastures, and driveways, and if dirt will have to be moved around the land before anything can begin.

However, it would also be a good idea to have a partial topological survey done as well by your surveyor at this time which would cost around $1000. Having this information from the start will not only tell you if the lay of the land is as you think it is, but it will give you a better idea of what will be involved in the site development before you start any construction. Another good idea is to go onto the county website and look at their aerial maps. You can find the parcel of land you are considering and click on various options to see the official information on your land including, most importantly, whether your land is right in the middle of an unobvious flood plain!

Next you should **"walk the land"** with your topological survey map in your left hand, the list of zoning restrictions in your front pocket, your rough drawing of where you want your buildings, pastures and driveways in your back pocket, a surveyor's measuring wheel in your right hand, and your realtor (if hopefully you have one) at your side carrying little survey stakes, a permanent marker, and a sledge hammer to **stake and label** important locations as you walk along the property. By doing this, you will know whether or not your building plans are realistic as far as placement goes. As I previously mentioned, moving dirt around to accommodate the construction is a major "hidden" expense unless you go out and walk your land and roughly determine what dirt might have to be moved. In fact, the need to rearrange God's natural plan for your property can require God's wallet to finance it and make the whole project just plain unaffordable. Remember too that **pole barns and arenas require perfectly flat land, unlike a house**. So please keep that in mind when you are walking your land in that the less dirt needed to be leveled, the less money you will have to spend. Also, be sure to mark potential places near the house location where a well can be dug. Our county regulations required at least 100 feet of separation between the well and a septic tank or field.

But let's say, and I surely hope it turns out this way, that the parcel of land is looking really good at this point and you are ready to buy it.

LESSON #11: **Before you actually purchase the land, be sure to put in the sales contract that the purchase of the land is contingent upon at least four things: (a) that the land actually does perk and that a septic permit can be issued to you in accordance with your desired use of the land;* (b) that the well has water, the well water is potable and uncontaminated, and the well has at least a minimal flow rate that I'll specify later; (c) that the contract warrants that there are no toxic waste dumps on the site; and (d) that there is access to electric power lines within a reasonable distance to serve your property (no more than about 2,000 feet from the utility pole to the house) .**

***Don't forget that if money is an issue, here is where you would also specify that the sand seam for a septic field cannot be found, as an example, more than about 6 feet below the surface of the land or the land purchase will become null and void.**

As previously mentioned, the seller should pay for a perk test if one has not yet been done. However, you will have to pay for a surveyor to mark the boundaries and do a topological map. Again, the reason it is important to stipulate that the purchase is contingent upon a septic permit being issued to you "in accordance with your desired use of the land" has to do with three basic issues. The first issue has to do with the potential limits on the size of your approved system. Although your property may perk, remember that is just a starting point because it may "not perk enough" to accommodate a 3 1/2 bathroom house, plus a barn bathroom you intend to have. How many baths you can have must be determined by the county sanitarian and that will be that. So if you have a larger number of bathrooms in mind than your county sanitarian will allow, that could be a land purchase buster and therefore must be written into the contract. Second, the location of the septic drainage field could be a deal buster. If the location of the septic field makes the location of your buildings and pastures impossible, then the land won't work for you. Third, if the depth of the drainage field is 18 feet and your budget can't handle such an exchange of dirt for sand, then, again, you will have to find another piece of land.

Here is the <u>**septic permit "order of go."**</u>

1 - the seller has the land perked, and if it passes the perk test, then the seller should give you a county health department, certified, **septic feasibility report** stating the location required for the septic drainage field and the number of bathrooms the system can accommodate;

2 - now you go back to the county sanitarian with the survey of your proposed property, including the topological survey, and your sketch of the site plan and then the county sanitarian tells you whether or not you need an engineered field;

<u>**NOTE:**</u> After you have purchased the property, you can complete steps 3 thru 6.

3 - if you do not need an engineered field, you can go to an excavator and have him prepare the drawings for a septic system which you would take back to the county sanitarian for approval and the issuance of a septic permit;

4 - if you do need an engineered field, you must go to a licensed civil engineering firm which will prepare the drawings for an entire engineered septic system which you then take back to the county sanitarian for approval and issuance of a septic permit;

5 – the drawings for an engineered field are done on a site plan showing the location of all buildings including the house, the barns and the optional arena;

6 – if you didn't have an engineered septic field with a site plan done by a civil engineering firm, you may need to eventually have an official site plan drawn by a civil engineer showing the placement of everything on the land.

But now, in addition to the relatively minor costs you have incurred when considering your "chosen land," you should elect to incur the more significant cash investment of paying to have **a well** drilled in order to determine the water quality, if the water is potable, and the well's flow rate (about $3000 to drill to a depth of about 150 feet).

The obvious "little problem" about drilling a well at this point is that you have to figure out "where" to drill the well. Although you have a rough plan for where all your construction will occur, those building placements and sizes aren't "set in stone" at this point and things could change. But at the same time you don't want to buy a piece of land with all your hopes and dreams on the line and then find out there isn't a viable water source on the land. So you should drill a well, and you will have to make an educated "guess" as to where to drill it.

Then, on top of guessing where to drill the first well, you will have to be ready with more than one "guess" spot should the first, second or (Heaven forbid) the third drilling yield an unworkable well (and yes, you will have to pay for each potential well that is dug which will quickly increase the original $3000 outlay of cash for drilling just the first well).

But the good news is that you (hopefully) remembered to mark potential well drilling sites when you walked the land and marked those spots taking into consideration the local restrictions on well placement. What you don't want to have happen is for you to be standing there in the middle of what amounts to an empty field, with the guys from the well drilling company, and all their equipment, ready to drill, and find that you have to make a wild "guess" as to where the well should be drilled only to find out later that the well was drilled in the wrong place according to local ordinances for well placement. Here is where your topological map and your land walk pay for themselves a hundredfold in time, money and effort.

NOTE: Be sure to ask your county sanitarian in the County Health Department what the restrictions are for well placement (as well as the septic system). As mentioned, one restriction in our county is that a well cannot be dug closer than 100 feet of any part of the septic system such as the tanks or the field. Where water is actually found can therefore, in the end, obviously affect all the building and pasture placements and cause the need for a planning re-think and re-do.

You will need a permit in order to dig a well. The permit is obtained from the county sanitarian. Once you obtain the permit from the county sanitarian, you take the permit to the well drilling company of your choice and give them the permit and set a date to have the well drillers come out to your property and drill the well. Of course, it is essential that you be there!

The well driller will drill down until he finds what he considers to be an appropriate source of water. In our state and county, the well driller would be looking for clear water with a minimal flow rate of 4 gallons per minute. Such a minimal flow rate, however, would not be sufficient for your practical water usage needs. In order to flush toilets and take showers, you will need a flow rate of at least 25 gallons per minute. Therefore, if the well driller comes up with a low flow rate in your well, then you will need to have a very large external water storage tank installed on your property and have your well continuously pump water into the storage tank to build up enough water pressure for your well to be realistically functional.

Your right to have a water storage tank on your property may or may not be possible due to local ordinances which you would have to determine. Do not assume that you can have a water storage tank on your property or that you can have any kind of water storage tank you want or have a water storage tank placed in any location you want. You must be sure to check out all the applicable ordinances should a water storage tank be required. It would also be important for you to find out the costs of nearly continuously running a well pump which may be too expensive for your operating budget, and therefore be a deal breaker.

Once the well driller has drilled what he believes to be a satisfactory well, he will take water samples from the well. The well driller will take the water samples to the county sanitarian for testing to establish the water's purity and to make sure there are no toxic pollutants in your well water. If the water samples pass, then the county sanitarian will record that the well has passed inspection on his records regarding your property.

If, however, the well does not pass inspection, then you will have to decide if you want to pay to drill another well or find a different property on which to build.

Remember that there is also such a thing as a dry well. Not all land necessarily has water on it or water that is both drinkable (potable), and unpolluted. Realistically, knowing whether or not you have water and that it is potable, etc. is too important an issue to ignore which makes the need for this upfront cash outlay and property purchase stipulation imperative.

Chapter Three

After You Purchase the Land – Site Development

So now you are at the point where everything required for site development has checked out, you received approval from your attorney, and you have purchased the land, and you have the deed to the land, and the deed is registered with the County Clerk.

Time to celebrate!

Okay—celebration over. You've got a lot of work left to do!

It's time to begin doing a thing called **site development**.

LESSON #12: Now you have to firm up your rough site plan of where you want all your structures to be built on your land, the sizes of your structures and your pastures, and where all your access roads and paths will be. You will then have to take your rough drawings to a civil engineering firm.

Here is your next big cash outlay. This interaction can cost as little as about $5000 and get as expensive as $20,000 or more, and there is no way to avoid spending this money.

- First of all, a licensed **civil engineering site plan and septic plan** will be required if you need an engineered septic field in order to eventually be able to obtain a building permit and a septic permit. If you don't need an engineered septic field, and you don't need a civil engineered site plan, you will still have to draw a site plan yourself, and you will need to hire an excavator to develop and draw a plan for your septic system.

- Secondly, the advantage of hiring a licensed civil engineer to do the site plan is that he will make sure that the placement of all infrastructure and buildings on your construction site are in compliance with the local ordinances, and that drainage will operate properly.

- Third, the engineer will make sure that all your buildings "fit" on your land as they comply with the local ordinances. (Here is where you might be disappointed to find out that your well was dug in the wrong place.)

- Fourth, the site plan is the document that your excavator will use when he prepares your site for construction.

- Fifth, if you give him permission, the civil engineer will determine "what dirt" has to be moved "where" in order to accommodate your construction. This is where your bill for services can increase from a few thousand dollars to several thousand dollars. Why? Well, if your land requires a lot of dirt to be moved around, then the civil engineer will have to spend a lot of time calculating how much dirt has to be moved and where it has to be moved on your land, as well as potentially calculating how much dirt will have to be shipped into the property or off your property. This is how the bill could suddenly go from reasonable to exorbitant.

NOTE: It would be a good idea to give the civil engineer a "heads up" that you want to be informed BEFORE he starts making any "dirt" calculations if there are some required, and to give you an estimate on what his services would cost should he be given permission by you to continue with such calculations. Get a firm estimate in writing and specify that the end cost for services cannot exceed 10% of the written, firm estimate. This is an important written procedure that should be followed for all your contracts.

Although you could elect to pay to have the civil engineer stake the building sites for you on your land, there are two less expensive ways to get the job done. The first, and obviously cheapest, would be for you to site and stake the building sites yourself using the engineered drawing. The second is to hire back the surveyor to stake the buildings sites, et al, for you which would incur a more reasonable fee than hiring the civil engineer to do it.

So now that you have your survey and engineered plans done and you've staked your buildings and driveways, it's time to get to work and build that house and barn, put up a fence, load up the horses, and move in, right?

Well, that certainly would be a happy thought, wouldn't it? But, no, unfortunately, it's not quite time for that.

What it is time for you to do is go to the county tax office and get a **tax id number**.

That sure sounds strange, doesn't it? Why would you get a tax id number when you really don't have anything to tax? Well, nothing to tax other than the land.

Well, here's the situation at this point. Even though you now own the land, and know what you want to build and where you want to build it on your land, before you can actually start building any part of it, you will have to have a building permit. However, in order to eventually get to the point where you can have a building permit, you will have to start the process by having an address. But in order to have an address, you will have to first have a tax id number.

LESSON #13: There is a legal process of several steps that you must go through in order to obtain a building permit before you can build anything on your land.

Yes, there are **a number of steps involved in getting a building permit** which are not necessarily easy to find out about or accomplish. The problem is that even the local officials in most small towns aren't sure any more of where to go and get things done because ordinarily they don't "have to know" about such things anymore because "developers and builders usually take care of those kinds of things" nowadays. But when building a farm, you will most probably be acting as your own site developer and general contractor, and, as such, you will have to take on the responsibility of getting the process of steps done while at the same time most likely finding it quite challenging to actually accomplish the steps that will finally yield you a building permit. Our experience can at least give you a feel for an order of go. Determining where these various offices are located in your area, and then having to deal with each of the "officials," will be enough of a time consuming challenge for you.

Building Permit Acquisition Order of Go:

1-- take your deed to the county tax office to get a tax id number;

2-- take your tax id number to the local (village, town, township, city or other) zoning official and have him assign you an address;

3-- take your site plan (either drawn by you or the civil engineering firm) plus a septic plan done by the civil engineering firm or your excavator, and the septic feasibility report to the local zoning official for his approval of your site plan;

4-- take the site plan plus the septic drawings including the septic feasibility report to the county sanitarian for his septic approval. The sanitarian will approve the drawings which mark the locations for the septic storage tanks, the septic drainage field, as well as the connector lines that run between the tanks and the field. The sanitarian will then issue you a septic permit. This permit with the approved site plan showing the complete septic system is what you will give your excavator to dig the corresponding holes and trenches for the septic system;

5 -- now you can hire an excavator (if you haven't already) and have him construct your septic field;

6 - with your deed, site plan, and septic permit, you will next need to hire one or more builders and possibly an architect who in some combination will produce the required construction drawings for all your buildings.

7—the next step for you and your builder(s) is to obtain building permits. For us that meant we needed to obtain and fill out construction permits issued by the township local zoning official using forms entitled "Application for Zoning Compliance Permit" with attachments which were the septic permit and the local zoning official "approved" site plan. Then we had to take the approved construction permits and all our construction drawings to our area building authority that was a separate, independent organization (not part of either the township or the county) that would be doing all building inspections during construction. Depending upon state law, you may not need a building permit for building "just" barns and arenas and other farm outbuildings;

Summary for clarification: If you need an engineered field, then you MUST have had it drawn by a licensed civil engineering firm (who will also have to do a site plan for you regardless of whether or not you require an engineered field). However, if you do NOT need an engineered field, you can go to an excavator and have the excavator draw the proposed septic system on the site plan that you or the civil engineering firm created (but the excavator proposed system would still have to first be approved by the county sanitarian before the county sanitarian will issue you a septic permit).

The excavator will make a first pass on grading the general lay of the land, including placement of culverts to allow for the flow of runoff water across the land, the leveling of the building sites (as well as the heights of the building sites), the roll of your pastures, as well as putting in your basic construction driveway and your basic septic system.

Your excavator will remove the topsoil from the building sites down to the subsoil. Buildings cannot be constructed on topsoil in that topsoil is an unstable soil. In addition, when topsoil is mixed with water, it turns into incredible muck—the kind of muck that pulls your rubber boots right off your feet when you try and walk across it.

LESSON #14: Some people who operate bulldozers tend to act like bulldozers themselves in nearly every aspect—size, weight, and personality. Don't be surprised if you find that some excavators have a tendency to consistently underestimate costs and then overcharge you upon project completion. In our experience, some of them will also tend to be happy to do a lousy job and not comply with the contract.

But before you even hire an excavator it is important to remember that a farm is like a town with a road system and walkways. So it's important that we spend a minute here to take **another look at the road system** on your farm beyond the main driveway you've planned on your survey map.

First of all, let's consider how long a driveway you've planned. Yes, it's nice to have a long drive into your farm. In fact, it's inspirational. A long drive lets you roll along through the beautiful pasture land, under trees you plan to plant that will someday arch across the whole path, and a long drive will give you plenty of time to admire the entire peace and beauty of the wonderful farm you've built as you drive along it.

However, the practicality of it is that driveways are surprisingly expensive. And I'm not even talking about asphalt. I'm just talking about your basic driveway with a stone base and a gravel top finish. And don't think you can get away with doing a "cheap" base or anything of the like. Remember, there will be a lot of big, heavy, construction vehicles driving down that driveway during construction, and then big, heavy semi's (they can weigh up to 80,000 lbs) and garbage trucks will use it every week once you've moved in. So, if you don't have a firm base, and a good driveway, you will have ruts—unending ruts—hard-to-impossible-to-fill-in ruts, plus ponding (which may be "just" puddles in the warm weather, but will turn into slippery, treacherous ice ponds in winter), and other water flow problems. There will also most likely be places along your driveway where you will need to put in metal culverts (steel pipe or corrugated steel pipe at least 8 to 12 inches in diameter) under the driveway base to allow for the "natural" flow of water across your land, but redirected to flow under your driveway rather than over your driveway. The metal culverts must be installed right at the start of the first construction driveway. Otherwise your driveway will have to be dug out later to install culverts which will make a mess of your driveway.

NOTE: I strongly recommend that you have a gravel driveway, rather than an asphalt driveway. The biggest advantage to having a gravel driveway is that it can be easily "repaired." Driveways have a tendency to sink and settle or bump up in places depending on the season which causes "new" low spots and potholes to occur. If you have an asphalt driveway, the driveway will just crack and then crumble over time and cost you a lot of money for maintenance and repair, and potentially replacement. But a gravel driveway just needs to be dragged by a box blade behind your tractor (when the soil is unfrozen and preferably after it's been softened by rain) to flatten out the driveway again. Severe damage to a gravel driveway just requires a stone delivery and subsequent spreading of the stone. In addition, gravel driveways are not as slippery as asphalt driveways for vehicles, people or horses to either walk or drive across.

In locations where freezing occurs, your driveway should consist of at least 4 to 6 inches of gravel base (#3 crushed limestone gravel which is also referred to as 3 inch cracked stone, gray color, driveway base), topped by 2 inches of crushed limestone (21 AA which is a mixture of crushed limestone plus pieces of stone from one inch in diameter down to "dust"). The driveway should be a minimum of 10 to 12 feet wide, with at least 14 feet of width on the curves, and have a 10 foot by 20 foot wide parking area allowance for each car. If there are any 90 degree turns along the driveway, then the width of the driveway should be at least 27 feet at that point. Any "uphill" (and corresponding "downhill") areas of the driveway should have a grade of less than 15 degrees or it will be too steep. As part of your driveway, you need to plan in a turnaround for large trucks which is best accomplished by planning in a K-turn or a turnaround loop with a stretch of widened driveway and enough turning radius. A turnaround loop would require about a 100 foot radius.

For us, a gravel driveway cost about $1 per square foot, installed. That implies that a 1,000 foot long driveway (which is not unusual for a "long" drive back to your farm buildings and house) would cost you about $10,000 for a 10 foot wide driveway. Remember that we are only talking about the main driveway to the construction site and not the complete road system.

Our first excavator came recommended to us by the realtor who sold us the land. Considering, in the first place, how uneasy the realtor's honesty factor made me feel, I should have known better than to have taken any recommendation from him. But when you are new to an area, you don't know anyone and you figure that taking recommendations from anyone local should ordinarily reduce the odds of things going totally wrong, and such recommendations usually do. However, in our case . . . ,

We met with the excavator one summer day on our newly purchased land, along with the realtor, and the neighbor who sold us the land (the one who initially wouldn't move her driveway off our land even though it diagonally split most of our 10 acres in half—by the way, because of the expense!). We found ourselves doing our driveway planning standing between a thick mass of overgrown weeds on one side (part of our purchased 10 acres), and neatly planted rows of soy beans on the other side (part of our purchased land plus our leased 10 acres). The comparison of the wild weed growth on the one side versus the planted, structured growth on the other side was dramatic enough to take note. Nevertheless, everyone but me seemed to feel totally casual about walking freely along, over, and through the soy beans that some farmer from somewhere with someone's permission had clearly planted as a cash crop. Also of note, at least to me, was that our footprints were clearly not doing anything good for the bean plants.

The foot traffic issue over the bean plants was destructive enough. But then there came the point at which the neighbor and the excavator decided to drive across the bean field to envision where her new driveway would go in its new, more centered path, down between the two, ten acre parcels of our farmland. And, the neighbor and the excavator made more than one pass across the soybeans in their "envisioning" process as they subsequently crushed the soy bean plants under the excavator's truck tires. At this time, we realized that our lease contract hadn't properly defined our rights to specify where we would like our neighbor's driveway to be located - especially not so it would divide the 20 acres in half.

It was when the soy bean farmer suddenly showed up and started screaming at the neighbor for having destroyed part of his soy bean crop on the land he had leased from her--but was now actually our land, but I guess was technically still his in the sense that his crop was on it and he had leased it—that things got very uncomfortable.

Eventually, and perhaps inevitably, the scream fest scenario between the soy bean farmer and our seller flowed over to us as he apparently decided that we had become guilty simply by association. The whole event finally ended with the farmer pointing his finger at us with great authority and informing us not to ever bother trying to buy any hay from him because he wasn't about to sell us any. Even though we were essentially "innocent" bystanders, to his mind we were clearly accomplices to our neighbor's lack of foresight in serving her own purposes while disrespecting and destroying part of his crop. Okay.

Unfortunately I later learned that this farmer does in fact grow the best hay in the area, and to this day, he is not my hay supplier.

But, moving beyond the temporarily distracting issue of crushed soy beans and the understandably infuriated farmer, there was the new issue of which we had now become aware for a need to reduce the rapid rise angle of the hill at the entrance onto our planned driveway from the cul-de-sac. We had already recognized the need to diminish the tight 90 degree angled turn from the easement onto our actual property in order for the turn to our property to be gradual enough to maneuver the truck and horse trailer along the driveway regardless of the weather conditions, like slippery ice and snow. To our great dismay, as I already mentioned, this need meant we would have to use a large part of the adjoining neighbor's 30 foot northern half of the easement which we wanted to use as minimally as possible.

Trying to solve these two problems engaged a considerable amount of time and discussion on our part with the excavator. During these discussions, the excavator seemed grumpy and uncooperative about the whole thing which really didn't make any sense to me. As I stood there listening, I kept wondering why the excavator was making it seem like such a big deal to make the transition from the cul-de-sac to the higher ground of the entrance to our property be gradual? What exactly was the big deal? And why couldn't the curve be gradual and wide enough so that I could safely maneuver our horse trailer onto our property and others could safely drive onto our property in large vehicles such as semi tractor trailer trucks?

Why? There really was no good reason why. No reason why unless, as it slowly became obvious to us as we stood there listening to him, that the excavator was just looking for an excuse to do things his way to cut his costs on the first pass, and then eventually jack up the bill with extra billable hours of work when we complained and he had to "fix" things to be the way we wanted them.

Although we eventually had a firm contract with this excavator--which his original behavior warned us we better have—we still made a mistake. We made the mistake of not actually, physically staking the location of the driveway, which would make the location literal and unarguable. We only just verbally discussed the location of the driveway with the excavator on that original meeting day when we walked the site with him. Without having put actual wood stakes in the ground to precisely mark the driveway we wanted, this gave the excavator, to his mind, the freedom to do the driveway his way - cheap, steep, sharp turned, and totally not what we had discussed.

The excavator went and created an entrance onto the driveway from the cul-de-sac that was not only in the wrong location on the easement, but was so steep that upon entering it from the cul-de-sac, it was like being positioned on a rocket pad ready for launch. If you survived the launch, then you were faced with a subsequent turn onto the actual property that was so tight that we had trouble even getting our SUV to make the turn. In addition, the driveway he installed was so low in spots and so uneven, that it didn't even work as a construction driveway. What he was supposed to do was grade up and level out the land where the driveway was in order to form a basic profile which would form the correct base for the finished driveway which would then result in needing minimal finish stone and/or leveling at the end of the farm construction project.

Of course, when we complained, the excavator argued that he would "fix" the driveway on the final finish. But when the driveway isn't in the right location in the first place, and it hasn't been correctly graded in the second place, and thousands of dollars of stone have been put in the wrong place as well as on an inappropriately shallow base, it would result in an astounding expense to correct all these errors later on with stone. Bottom line was that this excavator was a louse and setting us up to gouge us for a lot more money in the end.

Yes, you guessed it. We had to hire a different excavator to correct the first excavator's lousy work. And that next excavator wasn't the last excavator we had to hire. As such, be warned not to make any contract with an excavator that requires you to pay any part of the contract in advance or at most anything more than a minimal percentage of the overall contract. And have specification drawings in contracts.

LESSON #15: Stake all driveways and walkways to be built. Consider a short driveway unless money is no object. Be sure that the excavator has committed to moving the dirt around to make your driveway as close as possible to what will be the finish grade all along the driveway's length and that he has put in any metal

culverts at low spots that are needed under the driveway. Also be advised that you must have the "special gray" PVC pipe put under the driveway for the electric lines if you will be having underground electrical service and your underground electrical wires will have to cross under your driveway. Make sure that the excavator is committed to making the turns, and everything else about the driveway, "exactly" how you want them by staking them, and then including drawings in your contract.

LESSON #16: Make sure that your contract with your excavator states that all the dirt that has been removed from the septic field will be redistributed on your property and according to your specifications. Also put a schedule with dates in the contract including the completion date of the redistribution of the dirt. Remember that soil, and especially top soil, is valuable and note in your contract that no soil is to be removed from your property unless agreed to in writing.

All building site topsoil has to be removed by your excavator before construction begins and placed in a pile on your land for later use as the final top dressing on your land. The other, non-topsoil dirt (subsoil) from excavated holes such as from the septic system and the house foundation holes, needs to be placed in piles near the various areas where you will want the subsoil to be distributed for fill later during the construction (for example, as back fill around foundations), and also once the building project is complete for aesthetics such as berms and for water run off swales.

You will need to determine a place in advance where the topsoil can be mounded for later use. Topsoil must be scraped off the building sites because it is a very unstable soil, and cannot be built upon. However, topsoil is very valuable and is needed as the top dressing layer of soil on your land for your landscaping and for your pastures and for your lawns. At the end of the construction project, the excavator will make one last rough grade of the property, and then do a final grade of the property using your topsoil pile. While the topsoil is being stored in a mound, it would be advisable to take some measure to prevent weeds from germinating in the topsoil pile by either covering the pile or putting down a weed growth preventative over the entire topsoil pile, and spraying existing weeds with a weed killer product.

Why?

Being an enormous pile of rich topsoil, ideal for quick and easy germination equivalent to a gigantic flower pot ready for planting, all the topsoil needs is an opportunistic weed to take hold. In our case it was thistles. Thistles are such pernicious weeds that there are actually laws (at least in our state) about controlling them. They are tall, nasty, "woody" stemmed, prickle covered weeds that are difficult to eliminate once they have taken hold. More importantly is that they are NOT what you want your horses grazing on in the pasture.

Now back to excavators. One excavator should be possible to hire under contract to do both the septic system and the main construction driveway. Meanwhile, you'll remember that our septic field needed to be dug out down to 18 feet. The field also has a length of 50 feet and a width of 20 feet. A hole 50 feet by 20 feet by 18 feet is a big hole with a lot of dirt to be removed and then replaced with sand. What we didn't anticipate and plan for was the excavator taking the subsoil dirt out of the hole and then just leaving it on the property in one, mountain high pile. That left us with a lot of dirt that we now had to do something about and that incurred a lot of additional, unexpected cost. The dirt from an 18 x 50 x 20 foot hole left us with an amount equivalent to about 10 railroad cars full of dirt that now needed to be moved around the property--but without a railroad track on which to move it!

True, we had plenty of places to use the dirt, as you will also have. But not having realized this issue in advance caused us a significant cost overrun by essentially requiring us to pay to move all that dirt twice.

NOTE: Make sure that your septic lines are not going to run under any later constructed buildings or driveways. This is another reason why it's best to have a civil engineered site plan which you've taken to the County Health Department to have the locations of the entire septic system in regard to your buildings and driveways, etc., approved. It is also why you must then stake out building placements on the actual land.

LESSON #17: Have the excavator dig a drainage pond if one is needed.

Once again, if water drainage and flow and the placement of the drainage pond are not obvious to you, but you feel the need to have a drainage pond, then it would be best to have a civil engineer determine water flow and where a drainage basin should be located. The civil engineer can also determine a good estimate for the appropriate size of the pond that needs to be dug. Be sure once again to have in the contract with the excavator that the distribution of the dirt from the pond hole will be done according to your specifications, and done according to your schedule.

NOTE: The word "schedule" is used here because firm dates are not workable. For example, you might want some of the soil moved to fill in around a foundation. But obviously the soil cannot be moved until the foundation is complete and has been inspected by the local building inspector. A firm date for that wouldn't be known. However, to protect yourself, you should have something in your contract to the effect that enough dirt would be moved from the stored pile to fill in around the foundation within 3 days of notification to the excavator by the owner that approval for the foundation has been given.

NOTE: You may have noted that I used the word "inspected" in regard to the house foundation. However, whereas houses have to have the building inspector come out and inspect and then approve or disapprove the building at certain specified points during the construction, farm buildings in some states do not require such inspection.

LESSON #18: In reality, since you are the site developer and general contractor get firm prices and not estimates and specify EXACTLY what you want done/supplied from all builders and other subcontractors that you hire directly. Don't be ambiguous or casual or overly trusting about anything. But if it happens that you can only get estimates, then write on the estimates that the final cost cannot exceed 10% of the total estimate in the final billing. If the builder or subcontractor will not agree to such, then find a different supplier because they are probably dishonest. Remember that it can actually happen that the original estimate for the work was too high, and that the actual bill in the end is lower than the estimate was. Honesty happens! But this only happens with "honest subcontractors"—which at some point in your project you will be forced to consider may just be an oxymoron, and that no such person exists.

Excavators will have three basic phases of engagement with you. Although you should be able to work with just the one, original excavator, again, don't be surprised if you get sufficiently angry and fire the original excavator at some point. In our case, we were forced to use four separate excavators.

Phase One with the excavator will basically have him bulldozing to:

(a) make a first pass at the general lay of the land the way you want it;

(b) level the pole barn building sites;

(c) dig the septic field hole, lay pipe, and replace it with sand; dig the septic storage tank hole and place the septic storage tanks; dig the trenches for the pipes that run between the septic storage tanks and the septic field and lay the pipes (but not the lines that run from the storage tanks to the house or barn which are done at the very end of construction);

(d) put in a construction driveway;

(e) dig a run off pond if needed.

Phase Two will have the excavator coming back to:

(a) dig out the house foundation if you are also building a house; the foundation is poured; then the excavator returns and backfills the house foundation;

(b) dig the trenches needed for utilities to run between the house and the barn and other outbuildings, as well as to any other locations where utilities are needed such as electrical lines out to entrance gates, electrical lines to pasture fencing for junction boxes, plumbing runs to waterers located in both the barn and in the pastures. Overall, the basic trenches are for septic, plumbing, water, electric, and natural or liquid propane gas.

NOTE: The excavator must have, and will require, that all areas for placement of the utility lines be marked by you and staked by you or one of your builders on the property before he can dig the trenches for the utilities.

NOTE: The septic lines that need to be trenched between the septic storage tanks and the septic field are indicated on your site map approved by the county sanitarian. The excavator will use that approved site map to locate the septic line trenches himself. The excavator will come back to dig the final connection trenches for your septic system that run between the house and the barn near the end of construction. Remember that no part of the septic lines, the storage tanks or the septic field can be covered by any buildings, driveways or any other type of compacting weight structure or activity.

NOTE: Almost all the utility lines can be trenched late into construction **EXCEPT** for any lines that will run under cement. Cemented areas would include at a minimum such things as your barn aisles, a driveway pad by your garage or barn, sidewalks, and patios. As such, things like automatic waterers and the drain line for the inside wash stall that will be in your barn, must be trenched in before any concrete is laid. However, **all utility lines must be trenched in after the rough electrical and plumbing is done in the barn.**

NOTE: Remember to also plan out and lay pipe under the construction driveway for utilities to avoid digging up the driveway if some of the utilities will be running under the driveways.

NOTE: All water lines and septic lines must be dug **BELOW** the frost line which in our state is about 48 inches in depth. Other utilities are put in trenches ranging from 2 to 3 feet in depth.

Phase Three will have the excavator:

(a) do the final grading and spreading of topsoil on the property;

(b) do a final driveway;

(c) put the footing in the arena.

By now I'm sure you have figured out that the cost of excavators in your overall project is HUGE! A rough estimate of the excavator cost is about 10% of the entire project.

So now you have your building sites at the heights you want them and they have been leveled where pole buildings will be built, and your septic system is in—and the drainage pond has been dug, if needed. Remember that if you are putting up pole barn structures, they require perfectly level land on which to be built, unlike a house. Piles of dirt from the septic system hole should be placed near where you think they will later be needed. You have the lay of your pasture land the way you want it. Plus you have a basic driveway that gives you and the construction people access to your land.

Celebrate! It's pretty darned exciting to drive down that drive, even though it doesn't actually go to anything but dirt piles and wooden stakes at the moment. Nevertheless, it's been a struggle just to get to this point, and at this point you can now begin to "see," even more than before, all your dreams of what will someday be there at the end of that driveway. So, celebrate!

Shocker #4: If you haven't figured it out on your own at this point, I will tell you that building a farm requires a lot of "upfront" cash beyond the purchase of the land.

Whereas contractors who are developing major commercial or residential sites can get small business loans and construction loans from the bank to do site development and general building construction, an individual consumer cannot get a small business loan or construction loan for a non-commercial project to just build a barn/outbuildings. Individual, non-commercial, consumers can only get a construction loan to build an individual residence—that means a people residence (a house) that can also include an animal residence (a barn/outbuildings) being built at the same time as the house. The cost of the outbuildings (which means barns and arenas and such) in a horse farm project that is non commercial must be finessed into the size of the building loan for the house or else cash will be required for the entire "outbuildings" project.

Let's get back to driveways and such. Again, it is important to note that there are two phases to driveway construction and there are some factors regarding the driveway that are involved with utility installation. You will remember that both the placement of your driveways and the type and placement of **utilities** must be determined in your original site plan discussions and noted on your site plan. Although there may be necessary refinements as you receive feedback information from your utility providers, you still must have a plan from the beginning that you can modify.

The utilities concerned include:

- -- water
- -- electric
- -- natural gas or liquid propane gas
- -- septic
- -- telephone
- -- cable or satellite for TV
- -- broadband internet

In addition, you should be aware that the responsibility of the utility provider is simply to run the utility line to your house. Then, in turn, it is the responsibility of the owner to run the utilities from the house to any other buildings or sites on the property where the service is needed. It is also the responsibility of the owner to run the utilities within the buildings.

<u>LESSON #19 :</u> This means that you must be sure to run underground pipe between your house and the barn that allows for telecommunication wires, sewage pipes, gas line pipes, and plumbing pipes for softened and unsoftened water to be run through the pipe. Again, all utilities should be trenched in right after the rough electrical and plumbing are done in the barn and before any cement is laid.

(1) <u>water:</u> You have already "guessed" and located your well and the well has been drilled. You will still have to determine where you want and/or need access to water on your land and in your buildings, and then trench the water lines to those sites. There are many places where water will be needed both inside and outside. As an example, you will need exterior freeze proof pumps on the property near the hay barn and near the pastures, as well as in the barn aisle. If you are going to have automatic waterers, you will need to trench water lines to them in each of the stalls and in each of the pastures. You will need water in the barn as well for the wash rack, the bathroom, and the office, and potentially for a water-based heating system. Again, all water lines must be staked and marked by you for your excavator to trench and then the plumber to lay the pipes. The plumber will make the final plumbing connections near the end of construction.

<u>NOTE:</u> Remember that freeze proof pump lines must be trenched in the barn before any cement is laid.

(2) <u>electric:</u> Electricity is a basic utility that needs to be trenched in at the beginning of the construction project (after the basic excavation is completed and the house foundation is in). The first thing you need to determine is whether you want to have underground electric run to your house or above ground wires. Although it is cheaper to have above ground utility poles run into your property, above ground wires are not only unattractive, they are potentially dangerous if they should fall. Above ground wires are also more susceptible to ice and wind damage that can leave you without electricity while the wires are being repaired. And, yes, you will have to pay for either type of electric power installation you decide to have on your property—above ground or below ground. The electric company will trench in the underground wires to your house using their own equipment. Then it is up to you to stake and mark where all electrical lines must be laid to be distributed to all other sites (other than your house) on your property, and up to you to have your excavator dig the trenches for those wires. Your electrician will lay the lines in those trenches and make the final power connections near the end of construction.

NOTE: Installing an automatic switch over, natural gas or propane gas-fired generator on your farm is a good idea to provide automatic back up electrical service via gas should you have a power failure. This will cost you about $6000, installed. Planning for either an automatic generator or a manual generator will also require that you have two electrical panels in your house. One "stand by" panel provides a connection to all devices such as the well pump, garage door openers, computer networking equipment, television, lighting and heating furnace that will operate when you switch over to the generator. The "stand by" panel will feed electricity from the generator through a transfer switch to all devices that will have electrical service during a power failure. The other electrical panel provides a connection to all other devices that will not be powered by the backup generator such as an air conditioning compressor and self-cleaning, electric stove ovens.

NOTE: Don't forget that if you opt for underground electric which requires wires to be run under your driveway, then you must be sure to have the excavator put in the "special PVC pipe" needed for electrical wires under the original, basic driveway to accommodate the wires.

The process for getting your electric company to connect you to the electrical grid through new wires and a transformer begins by you calling the electric company and telling them you want to have electric service installed on your property.

The most important thing to remember to tell your electric company is that you want 400 amps of electrical service run to your property.

The ordinary service run into a property for a house is rated at 200 amps. This amperage will not be enough to provide the electricity needed for both your house and outbuildings (barns and arenas). If you have bought or own a property with an existing house on it, and are planning to now build barns and arenas, then you will need to pay to have an additional service line run into your property rated at 200 amps from your transformer and have an additional 200 amp panel installed in your house to run service to the outbuildings.

In some states, including the one in which I live, you do not need to have an inspection in agricultural buildings to verify that you meet code in order to run electricity to outbuildings. You can theoretically just have an electrician run the right size wires for a single 20 amp circuit at 120 volts from your 200 amp house panel to your outbuildings. However, you will only be able to have minimal power to any building such as a few overhead light fixtures. In order to have fully lighted barns, proper arena lighting, heat, air conditioning, a gas fired hot water heater, and heaters in your automatic waterers, you will need to run at least 100 amps of a 240 volt service over to the barns and arenas to a distribution panel that contains switched protection breakers for new circuits. Obviously the normal 200 amp electrical panel in the house cannot handle such an additional power load in that the normal electrical demands of just the house will use up most or all of the basic 200 amp electrical service.

After calling the electric company and telling them you want to have a 400 amp electric service installed on your property, the electric company in turn will ask you to mail them a **copy** of your deed, your address (obtained with a tax id), plus a **copy** of your site plan with the electric lines marked on it where you "think" they should run across your property to your house. The electric company will then come out to your property and first determine where the closest utility pole is to your land. They will next determine if the electrical wiring line run you have noted on your site plan is feasible. Once the location of the electric line has been determined as feasible, and the electric company has "signed off" on the placement of the electric lines, then the wiring can be trenched underground or run in on poles by the electric company according to your choice.

For us, the cost to trench in the electrical service was about $12 per foot. Although per foot pricing seems reasonable at first glance, it adds up very quickly to a big number if your building site is set well back from the pole supporting the main electrical utility line. The electric company will install a temporary meter (usually on a piece of raw wood board) on your site for the construction phase if you are building the house and barn/outbuildings concurrently. Once the house has been framed, the exterior siding and roof are on, and the electrical wiring is in the house, the electric company will return and install a permanent meter on the exterior of the house. If you have an existing house, then the electric company will install a second, 200 amp electric service connected to the service entry point on your house.

(3) natural gas or liquid propane: If you are going to have a natural gas line run into your property, you will need to have that trenched as well assuming that you have that option. In our case, the cost to have a natural gas line trenched into our property was estimated at $20 per foot which would have cost us a total of about $30,000 for the trenching. Of course, your ability to have natural gas on your property will depend upon whether or not there is a natural gas main running along the main road off of which your farm is situated. You can't get natural gas trenched into your property unless there is a gas main on the main road or another main nearby. Generally speaking, farm country does not have gas mains running along the main road. However, if there is natural gas available, and you choose to have it, the local gas company will be the one who trenches in the gas line—not your excavator. If you are going to have liquid propane gas, then you only have to trench lines from your onsite liquid propane storage tank to your buildings where you want gas, and you don't have to trench in those lines until the rough plumbing and electrical are done.

Gas lines between the house and outbuildings, if you have public service natural gas or propane from the liquid propane tank to the house and outbuildings, can be laid in trenches by plumbers who are licensed to do it. However, the final connections to the actual gas source have to be done by the authorized service people from either the propane gas company or the public service natural gas company. You want to make sure to plan ahead and dig one trench between the house and the outbuildings that will contain all the utility service lines for electricity, water, communications and gas because there generally will not be enough room on your property to dig several trenches, i.e. one for each utility. (However, the gas line into your property from the gas main on the main road, if you are using public service gas, has to be run in its own trench.)

Also, your decision about having either type of gas service—natural or liquid propane--will depend on whether you will want gas as an energy source for: heating (both furnace and hot water), cooking, and/or clothes drying. If you don't want gas for any of the preceding, then you won't have to bother with this section.

We found our natural gas company to be totally uncooperative. We couldn't get anyone to give us a viable quote on the general cost of installation over the telephone. In fact, we couldn't even get anyone to return our calls to schedule an estimate for installation of a gas line on our property. The "field" people made themselves totally unavailable to us. So even though the natural gas lines ran right past the private road leading to our property about 1500 feet from the location of our house, we eventually had to give up on having a natural gas line run into our property.

Why? Well, the consistent excuse we were given was that the installers and estimators were too busy doing "big" construction projects and didn't have time to get back to us. In other words, they couldn't be bothered with a little, individual customer. They were only interested in dealing with large, commercial developers. Somehow, by the time you hear that excuse about twenty or thirty times when you call, your mind just can't help but think of the word "kick backs." Curiously enough, even though the electric company and the gas company were one and the same in our area, even the electric company employees told us that the gas company people were difficult and uncooperative. Also curious was why only one trench couldn't be dug to serve both the gas and the electric lines coming into the property. The question remains to this day in our minds as to why a double effort and double expense would have to be incurred.

So, as I said, we eventually gave up and opted for liquid propane gas. Yes, choosing for liquid propane gas meant we would be dealing with that "ugly" big tank you see sitting on almost every property that is a little distance "outside" of town, and has served forever as a defining feature on just about every farm. However, the good news is that nowadays liquid propane tanks with a capacity of 1000 gallons can be buried underground because they have developed a non corrosive coating for the steel tank that allows it to be buried. So you no longer have to have the ugly above ground, liquid propane tank staring at you. Of course, as with everything, you do have to pay the liquid propane gas supplier more money to have the hole dug to bury the tank!

Photo #11: Buried Liquid Propane Tank

Meanwhile, I haven't found any significant difference in use between natural gas I've had in homes in the past and my current use of liquid propane gas. There is the downside with liquid propane gas, however, in being dependent upon delivery which in turn means you can run out of gas if your estimated delivery by the liquid propane gas company doesn't match your consumption rate and you've forgotten to check your tank gauge to see if you need to call for an unscheduled delivery. Although running out of gas is not a big deal as such, it can be inconvenient and annoying. And, yes, I've run out of gas!

(4) septic: As you know, the septic system is one of the first things that must be dug and installed. Because the septic topic was covered earlier, I won't go into any detail here except for one important note. First, for general understanding, there are two steps in the septic waste removal process. First there is the plumbing that flushes the waste into the buried septic storage tanks which is done by gravity feed. Then the waste is transferred from the septic storage tanks out to a septic drainage field. This transfer of the waste from the septic storage tanks to the septic drainage field is done either by gravity feed or by being pumped out to the septic drainage field (an engineered field).

As such, it is essential for you to realize that all buildings must be set higher on your land than the septic storage tanks in order for a gravity feed to occur from the plumbing fixtures in the buildings to the septic storage tanks. An engineered field only pumps the sewage from the storage tanks to the septic drainage field. An engineered field has nothing to do with the lines that run between the plumbing in the buildings and the septic storage tanks. This is yet another critical reason why all your buildings should be set on the highest points on your land so that gravity feed can occur between the building plumbing and the septic storage tanks.

(5) telephone and broadband: Thanks to the "**universal service and access rights**" in the Federal Communications Act of 1934 (then made explicit in the Federal Telecommunications Act of 1996) that dictated that everyone, no matter where they lived, must be provided with service and access to the telephone network, the phone company was a pleasure to deal with for basic telephone land line installation. They have no problem scheduling the installation of your phone lines, no problem installing them (they even have a boring drill so that they can trench right under an existing driveway without disturbing it!), and the whole installation is free of charge! That's quite a difference from the preceding group of utilities.

All you have to do to have your phone lines installed is call the phone company and schedule the phone line installation and the phone lines will be installed. Phone lines are put in near the end of construction. Remember that you still probably need to have land lines in your house even though everyone in your family might have a cell phone. Land lines are still needed for your FAX machine as a minimum as well as potentially for your broadband internet access using DSL. Of course, technology is always advancing and landlines for these needs may eventually be unnecessary.

Getting broadband service to the farm was a totally different experience. Universal service rights don't apply for DSL which was reclassified from a telecommunication service to an information service, which is treated differently under the 1996 Act. Getting broadband service was difficult for us. Unfortunately, there are no universal service rights for broadband communications in the United States that provide a high speed connection to the Internet through an Internet service provider.

(6) cable or satellite for television: It's unlikely that you will be able to get cable in the country. Once again, it would also be expensive to have cable run into your property even if you could get it. As such, the satellite dish is the more affordable and simpler source for television service in the country. Yes, it's true that the wind and weather, as well as large animals (particularly deer on our farm) roaming or running past your dish, can affect your signal and interrupt your service. However, such episodes of interrupted transmission are infrequent and usually brief. But what is also true, based on my years of experience, is that even cable fails to deliver from time to time when the central server shuts down due to bad weather, widespread power failures, and during high demand shows such as the finals for American Idol. These cable server failures also tend to be longer in duration than a satellite shut down. So, there is no real reason to worry about not being able to have cable and "having to have" satellite. Satellite dishes go in at the very end of construction. However, you must remember to have the telecommunication pipe trenched and laid between the house and barn.

(7) <u>Broadband Internet:</u> There are three common ways to get broadband for your internet services: via cable, via satellite or via phone lines (DSL).

If you are able to get cable and want to pay to have it trenched in, then you will have to plan for this trenching on your site plan early on and coordinate the trenching with the installation of the electric and gas lines.

Satellite can also be the provider of your internet access. However, satellite internet has a double negative that doesn't add up to a positive in that it is currently both slow, and expensive. Nevertheless, you could find that satellite internet is your only choice for internet access.

Although your telephone land line with DSL would provide you with fast and cheap internet access, there can be a problem obtaining this service. In order to have internet access over your phone lines, your property currently must be **<u>within three miles of a central telephone service facility</u>** or the phone company will not provide DSL service to you. As such, again, satellite may be your only internet access source.

Even though we were within the three mile radius of a central telephone service facility, we encountered another "problem" in getting DSL service on our property. Once again, because we were "just" an individual consumer, the phone company didn't consider us "worthy" of their time. No matter how many times we called, the phone company couldn't be bothered to do the engineering calculation to determine if we were in fact within their three mile service radius (which we knew we were). In the end, we had to resort to writing to the state government (Public Utilities Commission) to complain about the phone company's lack of cooperation in order to finally have DSL installed for our internet access. Unbelievable! So, I will have to note here that even the phone company had certain departments for its services like the other utilities that weren't very user friendly. I guess I will have to modify my earlier enthusiasm and state that the phone company was only easy to deal with when it came to laying the land lines into our house, and that was probably only because they had to comply with the universal access law.

Please remember that any of the above information can change due to the constant technological changes of our times as well as the tendency for laws to be ever changing. Therefore, be sure to call and ask questions to get the latest information regarding the availability of utility services and their installation. Keep our experiences in the back of your mind simply as a guideline to enable you to get all the information you need regarding the services and utilities you will need on your farm.

Now the time has come to take yet another look at the "natural" state of your land.

Remember that this is going to be a horse farm and so the first thing you should ask yourself is what horses like to do. The obvious answer is, of course, that horses love to graze. So now, even before you think barn, it's time you had better take a minute and look more closely at that land you now own and see **what is actually growing on it** before you consider putting your horses out there to graze on it. Is there grass growing on it? If so, is it edible grass? Are you aware that there are plants, weeds, trees, and shrubs that are toxic to horses, including native plants to the area? Is your land free of plants, weeds, trees, and shrubs that are toxic to horses? What about rocks? Are there rocks on your land that can bruise your horse's feet when he goes galloping off across your pastures?

LESSON #20: All grasses are not equal. Many plants, weeds and shrubs are toxic to horses. Rocks have to be removed.

"What are you talking about?" you might ask. "Everyone knows that wild horses roam the plains and eat whatever grasses they find."

Photo #12: Unknown Mish Mash of Weed Growth

The key word here is "roam." Whereas wild horses can freely roam, domesticated horses don't get the chance to roam at will across miles and miles of open land and sort through what's available to find the best grasses, and avoid the toxic growth. No, the reality is that to be domesticated is to be confined. Therefore, the responsibility goes to you to make sure the grasses that are available for your horses' grazing are safe and nutritious.

NOTE: There are books that list which plants, bushes, grasses, and trees are toxic to horses. It would be a good idea for you to buy one of those books to make sure there's nothing "bad" for your horses growing in, near, or around your pastures.

So take a good hard look at that land you bought. Are you looking at a plethora of weeds including masses of those nasty thistles, and even perhaps the allergy causing Johnson grass growing along the edges of your property? Or are you looking at a thick mish mash of organic growth and basically have no clue as to what all is there?

Well, the bottom line is that unless the last person who owned your land was grazing horses on the land, the chances are that you had better consider getting rid of all the growth down to the bare dirt, and then seeding all the land yourself with a horse friendly grass mix.

You will need to kill off all the "natural" plant growth in the planned pastures preferably with a horse friendly herbicide purchased from your local mill or farm product supplier, but definitely a non-toxic herbicide. There is no other realistic way to clean out the weeds.

Photo #13: Prepared Pastureland for Seeding – All Natural Growth and Rocks Removed

But then, once you've removed all the "mystery" growth from your land, you need to consider what you should now plant for your **pasture grass**.

Well, you should plant pasture grass. No, I'm not being a "wise guy" here. What I mean by that is there are pre-mixed pasture grasses available at local mill stores and farm supply stores specifically blended for grazing horses. You'll notice that most of the premixed, horse-friendly, grass mixes say "endophyte free" on the bag due to the endophyte fescue scare when foals were dying soon after birth in Kentucky several years ago. The foal deaths were eventually linked to endophytes in the fescue grasses and why "endophyte free" grass is now used in horse pasture grass mixes. Do be sure to buy "endophyte free" pasture grass mixes!

NOTE: The premixed pasture grasses are mixed for the climate in your area. Therefore, a southern farm, pasture grass mix should not have the same pasture grass mix as a northern farm, pasture grass mix.

There are still a few more important things you need to consider regarding grasses. Please note that some of this is information that you should know about and should do, while some of this is information that you should know about, but should not do.

First of all, you do NOT want to plant grasses that make your pastures "bale-able" hay fields for your horses to graze on. That means you do NOT want to plant pastures with either just alfalfa or with an alfalfa/timothy grass mix. The reason why you don't want to plant such a mix is that your horse's diet would be too rich and would cause him/her weight gain, health problems, and to have too much energy. Leave the timothy/alfalfa mix grasses for the hay farmer to grow, cut, bale, and deliver to your barn for your horses to feed on in their stalls during feeding time. Generally speaking, your grasses out in your pastures need to be much lower in nutrition than alfalfa or alfalfa/timothy mix so that your horses can graze on the pasture grasses for many hours without upsetting his/her health or digestion.

The next thing to know is that the reason why Kentucky Blue grass is growing all over Kentucky horse pastures is because it's beautiful, hardy, and horse friendly. Then you should also know that there's "magic" in planting clover in your pastures. Not only do horses LOVE to eat clover, but clover also "magically" acts as a natural fertilizer to all the grasses growing in your pastures. Having clover in your pastures will keep your pastures rich, dark green, and healthy without having to add chemical fertilizers —that is, unless you allow overgrazing and/or trampling of the grass particularly in the muddy seasons.

So, okay, let's say you decide to plant a mixture of grasses like I did for health, beauty and self maintenance—something other than the basic, prepackaged, pasture mix. Here is what I used:

Kentucky blue grass	--	28 %	(5 lbs. per acre)
Smooth bromegrass	--	50 %	(9 lbs. per acre)
Orchard grass	--	17 %	(3 lbs. per acre)
Ladino clover	--	5 %	(1 lb. per acre)

So now that you've made your pasture grass decision, you call the landscaper, right?

__NOTE:__ Be sure that you have enough topsoil where you are planning your pastures to enable your pasture grass to grow.

__LESSON #21:__ In general, people who call themselves landscapers do gardens and lawns around houses, apartment houses, and shopping centers. Farms are not in their crosshairs. Nevertheless, ...

Surprise!

You're not even close to being ready to call in a landscaper.

But in the meantime, you have your site plan drawn complete with your house, barn, arena, outbuildings, pastures, fencing, driveways, and pathways, all with the proper setbacks and separations. You've had the surveyor or civil engineer (or you) stake the buildings and driveways on all the building sites. You've put in your construction driveway, septic, well, electrical lines, have done a rough first grade of your land, and killed off all the weed growth. So, if it isn't time to call a landscaper, then what is it time for?

Well, how about if you drive your car out to the end of your construction driveway and take another look out over your land? First take a minute to celebrate. Then after your celebratory moment, take another, closer look at your pastures now that you've killed off the weeds.

If you look at those areas that have been scraped bare by the bulldozers and then de-weeded with an herbicide, what you will most probably notice laying all over your pastureland, once your heart has gotten over how pretty the cleared brown land looks, is **__rocks__**. Lots and lots of rocks. In fact, it was amazing to me how I had never noticed just how many rocks are actually in fields. I'm sure, like me, that you've driven along through the countryside in the early spring and seen farmers out harrowing their fields getting them ready for planting, and thought to yourself how the dirt looks so nice and dark and rich and soft and ready to grow unobstructed whatever seed the farmer might decide to plant. It seems your whole mind set just naturally focuses on the "beauty," shall we say, and readiness of that soil. But while your eyes were busy looking over that turned soil, did they ever notice all those rocks also laying right there in front of you on that fertile, just-tilled soil? If you're honest, you'll probably have to say "No!" In fact, you'll probably find yourself asking, "What rocks?" There aren't any rocks! How could there be rocks?"

Well, the fact is that at least in the north, there are incredible amounts of rocks in the soil. I would recommend that the next time you drive past a farmer's field—in the seasons when there aren't any crops blocking your view of the soil—you take another closer look at that tilled soil and focus your eyes to look for rocks, and I'm quite sure that you will indeed see rocks!

The problem with rocks is that rocks and horses don't mix. Anyone who's been "in" horses knows that horses' hooves are easily bruised by rocks, and that bruised feet in turn means "down time" for your horse, and down time for your horse in turn means "no riding" for you. So the bottom line is that you have to get rid of all those rocks in the soil before they become pastures for your horses.

LESSON #22: First have the excavator remove all boulders as he is scraping and leveling the land or redistributing any soil or digging any holes or trenches. Then find "someone" who will probably call himself a commercial landscaper to remove the rocks from the pastures. It's important to find someone with the correct, specialized equipment so as to properly and economically remove the rocks.

There is commercial equipment that removes rocks from soil, but finding someone who has the equipment is not so easy. You could buy a "rock hog" and remove the rocks yourself using your tractor to pull the rock hog. But the question to ask yourself is what you are going to do with the "rock hog" once you're done removing rocks from your land. Plus, what would you do with the piles of rocks you've removed from the soil? Remember, not all rocks are created equal and some of those rocks are BIG. So, it's really best for you to look around for someone who will do the job for you—both remove the rocks from your soil and then remove the rocks from your land. Remember, however, that rocks are valuable for building retaining walls and landscaping. An honest subcontractor will give you credit on his labor for the rocks he removes and will subsequently use on other projects or sell.

One note here is that you might want to keep those rocks for yourself. You could just have the rocks piled up at the ends of the fields and move them around later for your own landscaping--if you're feeling strong or have the appropriate equipment to lift and move them, and also if you have the time. Having rocks around does come in handy. Rocks make nice edges around ponds. Big rocks next to farm signs and mailboxes protect them from vandalism. Rocks also make nice edges for driveways and can be great reminders to garbage trucks as to where the driveway edge ends and your grass begins to stop the garbage truck from running over your lawn! So keeping the rocks for your own use is a point to consider. The down side to rock piles is that they can be ugly if you aren't going to get around to doing something with them within a reasonable amount of time (a year or two) and can be both a "point of spook" for your horses and a perfect locale for a "rodent motel."

What's important to realize here is that your ordinary landscaper is NOT going to be the "man" for the rock removal job. But then who is? Basically all you can do is ask around and shop around to find that one person who is "the guy" who does rock removal, plus a final grade AND seeding of the pastures which "ordinary" landscapers either don't do or will charge WAY too much to do. Having this work done shouldn't cost that much either, especially if you allow the landscaper to take the rocks which he will resell TO the ordinary landscapers making your project a double profit for him. The cost should be around $500 per acre to remove rocks and seed pastures which is what we paid for our leased 10 acres to be de-rocked, prepared for seeding, and then seeded. In contrast, the "ordinary" landscaper who did our other approximately 6 acres of pasture charged us the "extra ordinary" price of $8500.

So now you've had your pastureland de-rocked and you are ready to seed those pastures, so you drive out to take one last look at those freshly prepared fields and look out across your land with pride and satisfaction and what do you see? Rocks! Lots and lots of rocks! What's up with that?

LESSON #23: One man's rock is another man's stone.

This is the beginning of what will be repeated over and over throughout the building process. The definition of how things should be for horses according to a "horseman" versus how things should be according to a "workman." When a landscaper looks at a rock in a field that is 2 or 3 inches in diameter, he categorizes it as an insignificant stone. Why? Because to him, he looks at the rock and sees that it is a size small enough for grass to grow up and around. But to you, the horseman, you see a rock that is big enough to bruise your horse's feet! Now try and explain the problem of bruised horse feet to someone who, not only doesn't know jack about horses, but really doesn't care for them or about them. It's like trying to communicate with someone wearing a blindfold and ear plugs. Of course, as with everything, the real truth is that the resistance is really all about having to do extra work, and then about trying to get more money out of you to do the work. Removing those smaller stones (as landscapers call them) will require more equipment and more time than the landscapers had put in their estimates, and they aren't about to do it. Plus, somehow, there doesn't seem to be a piece of equipment that exists to pull out those remaining "stones" which the landscaper will tell you means that they will have to be removed by hand.

LESSON #24: In the contract for rock removal, be sure to specify that all rocks greater than one inch in diameter are to be removed.

When I came out to the farm after the supposed rock removal, I found baseball sized "stones" still everywhere on the pasture land which then had to be hand removed. The need for hand removal of the "stones" resulted in a ridiculously large bill being sent to me for the additional work (about an additional $7000). You should know that I refused to pay the bill!

Remember we had talked about the pasture grass seed mix earlier. For review, you can either plant a prepackaged horse friendly, endophyte free pasture grass mix, or customize one. Remember also that in my customized grass mix that I used a heavy dose of Kentucky blue grass and added in some ladino clover. As I mentioned, the Kentucky blue grass is hardy and pretty while the ladino clover is horse friendly and provides natural fertilizer for the grass. Also remember that you DO NOT want to plant a "hay field" like pasture with either just alfalfa or a mix of alfalfa and timothy or it will be too rich for your horses to graze on. Also important to consider is the climate in which you live. There are some grasses that do better in the dry seasons (or climates), ones that over winter better, ones that tolerate moisture better, etc. That is the reason for planting a "mix" of grasses. Your horses will need and want something to graze on every day of the year regardless of the weather conditions. So, you have to have something growing in the pastures for them to graze on. If you can't figure out what to plant, then remember that you can just get the premixed seed at your local mill store that will be a correct mix for your climate. You can also call your local county agricultural agent and consult with him.

So now that the rock remover "landscaper" has done a final grade of your pastures and is ready to plant your seed, you must also consider that **planting the seed is "time sensitive"**. You already know that it's necessary to plant the seed at the right time of the year in order to maximize the probability of it successfully growing. In other words, you don't plant seed in the winter because it's too cold or in the summer because it's too hot and dry or, for that matter, in a windy season or a monsoon season, if you have one or the other of those, because seed can both blow away and wash away. Ideally, autumn is the best time to plant seed; spring being second choice in temperate climates.

Of course, being in accordance with Murphy's Law, our seed was FINALLY put down in July which is the hottest and driest month of the year where we live. Yipes. I'm still amazed to this day that we even have pasture grass. I'd have to say that God smiled down on us and gave us just enough rain over that first summer of seed planting so that the seed could take hold. As I stood there staring at the seed baking in the July and August sun, all I could think of was the thousands of dollars the final grading and planting had cost. It certainly didn't seem as though the seed would have much of a chance to grow or for the soil to stay in place should the seed not grow to stabilize the soil before the heavy rains in the fall, and the subsequent snow melts in the spring. But, thankfully, somehow it did.

LESSON #25: It takes two years for the pasture grass to have taken hold enough to withstand the abuse of grazing, galloping horses. Therefore, put the pastures in and seed them (season permitting) right after the septic system and the construction driveway are installed.

Although the first year's pasture grass growth is good, it is thinly spaced and doesn't have the kind of long, underground tendril root mats that make the grass rugged enough to withstand constant horse grazing and traffic. By the second year, the grass is quite strong, but it still needs limited, managed horse grazing. When the third season has come and gone, you'll have strong pasture grass. Of course, you'll always have to manage your pastures and the grazing, and be mindful of overgrazing and soil conditions—remembering that soft, wet ground plus horse traffic equals ruined pasture grass.

Photo #14: Derocked, Seeded Pastureland after Three Years Growth, Protective Fencing around Pasture Tree & Pasture Automatic Waterer

Pasture grass should be maintained at a height of about 6-8 inches. Shorter than that, your horses will tend to rip out the roots of the grass while grazing. Longer than that and the grass loses its nutritional value.

It would be nice to think that you will have the time to be able to keep your pastures right there in between those ideal height ranges, however, reality usually reveals itself to be quite different. The problem with the pasture grass growing too long, other than the nutritional loss, is that your cutter will make a mess of things when mowing. The grass will lay on the pastures in matted clumps of too-long grass that will choke off new grass growth in your pastures. More importantly, the long clippings and mats of grass can choke your horse. Horses look for easy grazing—just like kids look for easy snacking. They will gladly scoop up the dangerously long and matted clumps and suck them down oftentimes without enough chewing—or for that matter, without any chewing at all—and that could be a BIG problem.

So what do you do if you've let your pasture grass get too long?

You mow, mow, mow!

What do I mean? Take your lawn mower (a riding, zero turn, mulching blade, maximum blade width, mower) out into your pastures (rather than your tractor and cutter which do only a rough cut) and mow the long clumps that are laying on the pasture grass, and the pasture grass itself, with the mower set at its highest cutting level, which is usually about 4 ½ inches, every day for four to five days. That process will cut the long grass mats on the surface, and draw up the grass mats below the surface, then chop all the mats down to size, and eventually either cause the mats to rot and provide mulch for the grass or dry up on the surface and blow away.

If the mats and clumps are not excessively long at a length that would require repetitive mowing to shorten their length, and the clumps and mats aren't heavy and wet, then you can just blow the mats/clumps into piles, rake them up and cart them away. To do this, you would make circles with your lawn mower starting on the outside edge and working in toward the middle. You would mow in the direction that allows the blower part of your mower to blow the clippings continually in toward the middle. Once your "circle" gets small enough you would then rake the clippings into a pile, lift them into the back of your dumping Gator, and remove them from your pasture. You will quickly get a feel for how big a circle you can practically make before the clippings are too heavy for your mower to blow into a centralized pile for raking.

These systems work for removing excess clippings from your pastures, but I have to confess that it does take a lot of time to do that much mowing. Plus, technically, your pasture grass will be too short from the lawn mower mowing and the horses could pull out the grass roots and ruin the pasture. But that's really only an issue when you have too many horses grazing in your pastures. Clearly it's much easier to try and keep on top of maintaining that 6 inch grass height with the tractor and cutter!

Another benefit of keeping your pasture grasses cut in the correct height range is that it reduces the number of rodents and small mammals that love to "hide" in tall grass. The more trimmed up your pastures are, the less loaded they are with vermin, and then in turn, the fewer vermin you'll be likely to have in your house and barns, and, therefore, the less damage you'll have to your feed and structures, and on and on.

Photo #15: Fencing Protecting Pastureland

Okay, you've got the general lay of the land established, the septic and driveway et al are in, the pastures are de-rocked, seeded, and growing.

<u>LESSON #26:</u> It's time to put in the fences to protect the pastures.

<u>Photo #16: Fencing Protecting Pastureland</u>

I know it seems a little out of order to be putting up **<u>fences</u>** before anything else has gone up. However, that's just the point. Before anything goes "up" you will need to protect all the time, worry, and money you've spent on getting your pastures bulldozed, scraped, de-rocked, and then seeded. Workmen and deliverymen will soon begin driving onto your property once construction begins. There will be all sizes of trucks driving onto your property from pickup size to tractor trailer size. But the fascinating thing about these workmen and deliverymen is that they invariably take the most direct line to the building site regardless of the obvious existence of a driveway.

<u>Photo #17: Why You Need to Protect Your Pastureland</u>

It seems counterintuitive to me that one would choose to drive through foot deep mud to a work site rather than drive across a safe, easily navigable, gravel driveway. However, construction people consistently do it. They also have absolutely no regard for obviously leveled, prepared and seeded land. I mean NO regard. It's almost as if they are drawn to do damage because to their minds "to have a truck" is to "go off road." And, of course, they are summarily offended if you say anything or, frankly, find yourself resulting to swearing at them about it! Whatever. The bottom line is that even though you may not feel like thinking about fencing right at the moment, and all the decisions and shopping that go into it, you have to.

So let's talk fencing.

Ask yourself what "says" horse country more than long, neat rows of bright white fencing embracing endless fields of emerald green grass? I'm sure you will agree the obvious answer is "nothing." It's inarguably the signature look of horse country. In fact, it's so identifiable with horse country and all the positives that imagery represents that "the look" is found and repeated over and over in everything from housing developments to upscale shopping areas who all want to borrow from the beauty of such equestrian "horse farm" imagery. So how do you get that look on your horse farm in the safest, most economical, and attractive way?

Probably the thought of wood fencing, painted white, is the first thing that comes to your mind, and that was my first thought as well. However, as I started researching fencing, I immediately found a rather unanimous consensus among horse "authorities" and "celebrities" advertising wood alternatives, that wood fencing for horses, and for your pocketbook, is a poor choice. In the first place, wood is dangerous. If you've boarded at a farm with wood fencing you already know that horses easily kick through wood fences. Then, not only would a kicked in fence need repair, but potentially your horse would also need repair! The odds are that sooner or later a horse is going to kick through a wood fence in just the "right" way to split the wood and cut through his legs like a knife. In addition, the kicked out fencing makes an immediate "exit" door for the horse (and his pasture buddies!), which of course a horse is innately programmed to take as an exit door from the pasture--every time!

Then there's the issue of making and keeping the wood fencing white. White wood results from painting the wood white. Yes, I know that's pretty obvious, but what you have to consider is who is actually going to paint your fence white in the first place, and then who will be continually maintaining its whiteness over the years. This is no easy task, nor is it cheap in either material or labor costs.

Then, too, I've never seen a wood horse fence that has any real height to it, and horse fencing has to be high enough to (relatively) safely contain your horses.

The correct height for horse fencing is actually about five feet high (at least theoretically) in order to keep your horses from jumping out of their pastures. However, I'm not sure where you can get five foot "horse" fencing. My fencing is about four and a half feet in height, and, as I've mentioned, the horses have no problem jumping out of the pastures. I'm not sure that extra, hard to find, six inches would make much difference. However, getting "horse" fencing as high as possible is a good idea.

So for me, after doing an informational survey of fencing, I had three criteria I wanted to meet. I wanted: (1) white fencing; (2) high fencing; and (3) non-wood fencing. There are only two ways I know to get the white wood fencing look in non-wood fencing.

The first way is to use the PVC, stiff-rail, "plastic" fencing with a minimum of 4 rails. I've seen a lot of horse farms done in this fencing (usually only in 3 rail), and I have noticed several problems with it. First of all it looks very plastic especially when the sunlight is glinting off it. Second, it isn't really what I would call strong fencing in that it tends to lean both from environmental conditions such as heavy snow plowed into it, and from horses leaning against it. Third, the rails can be rather easily pushed out by a horse pressing against it.

The second way to achieve the look I wanted, and is the fencing I decided to use, is "flex fence" which is built using rolls of 5 and ¼ inch wide polyethylene plastic "tape" with 3 embedded metal wires. I built a 4 rail flex fence which has an approximate height of 4 and ½ feet when installed. The plastic "tape" is pulled tight with tension adjusters and strung through hangers that are attached to big, heavy, wood fence posts sunk deep into the ground and then set in place with cement. First of all, it looks really pretty. From a distance you wouldn't even consider that it's not wood fencing. In fact, even close up, your mind still doesn't really register that the rails are plastic. I think the reason for that is the massive wood posts do some kind of a trick on your mind that prevents your mind from "seeing" the plastic of the "tape" rails which subsequently results in making the "right" picture of wood fencing in your mind's eye, so to speak. I've been generally happy with the fencing. However, it's not perfect.

Photo #18: Pasture Fence Hot Wire Run Inside Top Rail of Flex Fence

I got the highest fencing the supplier sold which is the four "rail" fence which I've already mentioned when actually installed on the wood posts, results in a final fence height of about 4 and ½ feet. I also installed a "hot wire" which is a plastic coated wire through which high voltage electricity is run when its transformer junction box is plugged into an electrical source. The hot wire runs along the inside of the top rail of the fence, and is meant to keep the horses from leaning on the fencing. Therein lays two of the problems with the fencing.

First of all, have no doubt that your horses will lean on your fencing. When they do, it stretches the fencing. When this type of plastic tape fencing is stretched, it then has to be tightened. Tightening the fencing requires Superman-like strength as well as an expensive, long handle ratchet.

The second problem is that if you plug in your hot wire and use that to keep your horses from leaning on the fence, then you may find that your horses become "afraid" of the fencing, and in brilliant "horse logic" flow will then also conclude that they should be "afraid" of the pastures, and because they are afraid of the fence and the pastures that they should also in turn be afraid of the pasture gates. Such horse logic flow will next result in you having "problems" with turnouts, i.e. both in getting the horses to turn "out" into the pastures or to come back in "from" the pastures because they are now afraid to walk through the gate or past the fencing with its hotwire—even if you've turned it off—and even for several months after you've turned it off!

Nevertheless, the fencing I have hasn't caused any injuries, it's pretty, it's low maintenance, and you can retrain your horses to go through the pasture gates again even after they've been zapped by the hotwire which mine obviously have been. May I suggest pockets full of horse cookies as inspirational fodder to correct any hot wire issues that may have arisen?

UPDATE: An additional unforeseen problem has occurred with this type of fencing this year. We have had a prolonged period of drought which has caused the ground to shrink around some of the corner, cemented, wood, end posts and caused the posts to lean in toward the pasture from the tension of the plastic tape. The leaning posts have in turn caused the hot wire and the plastic tape of the fencing to lose their tension and gap. When we called the fence material supplier they told us that this was the fault of the installer who should have sunk the corner posts deep enough into the ground to prevent such shifting. The material supplier also said that all fencing corners should consist of posting with two, 45 degree angles and not just one 90 degree angle at any point along the fencing. Unfortunately our installer did do some 90 degree corners which seem to have added to the problem. Whereas the problem can be corrected, it incurs significant cost and inconvenience, and exactly who should pay for the fix needs to be determined! The "heads up" here for you is to make sure that every corner on your fencing consists of posting with two, 45 degree angles.

LESSON #27: The first step into fencing quotations for flex fence is deceptively low.

You will be quoted an amazingly low, and therefore appealing, price per foot for the fencing tape materials. But be forewarned that first price is only for the fence tape material. By the time you pay for the wood posts, the installation of the wood posts, the fencing tape materials, and the installation of the fencing tape materials onto the installed posts, the cost of the gates, and the installation of the gates, you'll find that the price has gone way up. Oh, and you will also have to pay to have the fence posts originally painted white, and then periodically (about every 4 years or so) you will have to re-paint the fence posts white—and when you do have to re-paint the fence posts, you'll also find how happy you are that you don't have to also paint any horizontal wood slats (as on a wood fence) in addition to those vertical posts!

NOTE: Curiously enough, 4 rail, flex fence materials plus installation cost approximately the same as the PVC 4 rail fencing materials plus installation. This installed cost is approximately $8 per foot of fencing. That means for 4,000 feet of fencing, it will cost you about $32,000 for either flex fence or PVC fencing. Therefore, cost is not the determinate on which to make a choice between these two types of fencing.

Also of note is that the flex fence material supplier is generally not the installer of the materials. The flex fence supplier should have authorized installers, however, to do the installation. Because there is a materials supplier, and then a materials installer, you should be aware that you will be billed by two different people. You will be billed by the fence tape materials people for the fence tape and the hot wire and hot wire junction box, and then you will be billed by the installer for the wood posts, the installation of the wood posts, the painting of the wood posts, the installation of the fence tape and hot wire and hot wire junction box, the cost of gates, and the installation of the gates.

There are cheaper ways to fence in your horses including wire fencing, barbed wire (which you should NEVER choose), and single strand, cloth or tape hot wires with some people just running a single wire along between stakes or posts. I personally haven't had any experience with wire fencing. What I do know is that wire fences are prone to cause injury to horses. I have seen a lot of horses with extensive scarring from wire fence "accidents".

I can share one story with you regarding single strand, wire fencing, however, that demonstrates that they do actually work.

A friend of mine who had to turn her horse out to pasture due to a show limiting physical condition was very concerned that the turnout pasture at the new facility had only one thin wire running around it for fencing. She told me that she was sure her thoroughbred would break into a gallop and run right through that wire. When she called me a few days after putting her horse out to pasture, she told me that sure enough, the first day she turned out her gelding with the herd, her horse started galloping like a wild man in his effort to establish himself within the pecking order. She told me she held her breath as she watched him gallop straight at that thin piece of perimeter wire along the back of the pasture feeling certain that her worst nightmare was about to come true. Yet, to her complete amazement, she said her horse knew EXACTLY where that wire was, and came to a screeching halt right in front of it. She said that after she witnessed that event, she never again worried about him getting loose from that pasture. However, I think it is rather obvious that such type of fencing does carry a lot of potential for both escape and injury and I personally would NEVER recommend it.

Remember that when planning your pastures, it is important that you **NOT** have a common fence line between your pastures so as to prevent "over the fence" fighting. A good separation distance between pastures is a minimum of about 25 feet. This spacing not only allows for the safety of your horses so they don't fight with each other over the fences, but it also allows you to easily walk horses out to the pasture between the fences without having the horses contained in the pasture able to reach over the fence and cause any distraction or trouble to the horse you are leading. The spacing also allows you to easily drive and maneuver your tractor both in between the pastures, and then into your pastures for mowing or for any other work you might have to do in the pastures. It also allows enough distance for the big trees truck to access your pastures to install your big trees.

Again it is important to remember that your pasture fencing should have angled corners at every corner. There should **not** be any 90 degree corners anywhere along your pasture fencing (of any type, be it flex fence or not) in which horses can become "trapped" by another horse or by a predator. Your horse should be able to run the pasture following what to his mind will be a continuous loop along the fencing because there are no right-angled corners to trap him in.

Before you call the fencing people, now would be a good time to pull out your site plan and look at the location of your pastures in regard to your barn and other buildings. In addition to determining the spacing between the pastures, the next thing you will have to determine is where you will want all the **points of access to your pastures**, i.e. the gates.

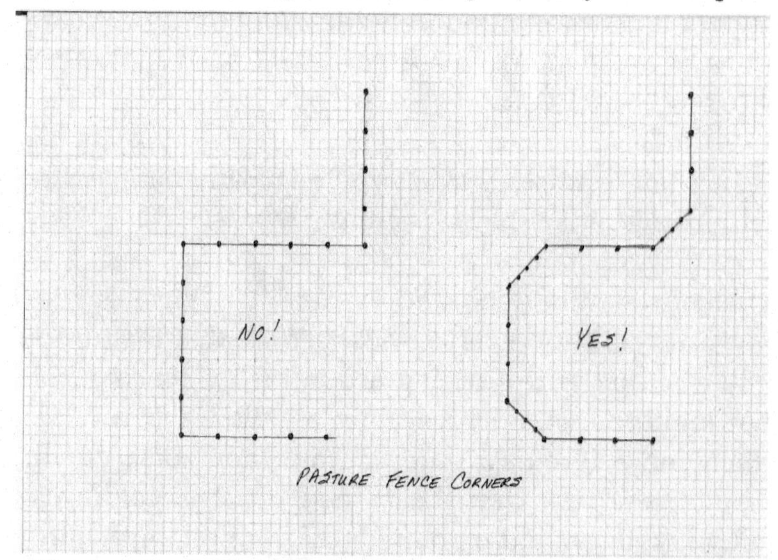

Diagram #1: Pasture Fence Corners at 90 Degree Angles (NO!) and 45 Degree Angles (YES!)

First you should consider that it would be best to have a short walk to the barn from the pasture, but not a "straight run" to the barn. In other words, it is best to have the entrance to your pastures be at a right angle to the entrance to your barn so that your horse doesn't see a straight line from the pasture gate to the barn. If a horse is panicked in any way and sees a straight line from an open pasture gate to an open barn door, you can rest assured he will beam onto that straight line, take it, and then pay absolutely NO attention to you! His "horse brains" will kick in and leave you in the dust. But, on the other hand, should the horse be in a panic, you will want to be able to get him into the barn on the shortest possible path. Therefore the gates should be at the end of the pasture nearest the barn, but not on a straight line run.

I insisted upon having two sizes of gates into each pasture. Whereas most people use an eight foot pasture gate to serve as both the horse and human entrance gate as well as the tractor gate, I just don't see the logic in that. First of all, gates tend to sag and they tend to sag almost immediately—but for sure they'll sag within a year of installation. A sagging, eight foot wide gate is impossible to agilely handle especially through mounds of snow and rock solid, slippery ice.

Photo #19: Horse/Man Pasture Entry Gate and Tractor Gate

The argument for the eight foot gate that I've been given is that you need that much room to "fit" both you and your horse together through the gate. Well, maybe if your horse walks sideways into the pasture you'd need an eight foot wide gate, but the last time I checked, there aren't many riding horses that are wider than about four feet when walking straight ahead.

Apparently another problem people have with their horses is being able to walk their horse through a "narrow" gate/area "with" them or "ahead of" them or "following" them. I suppose if that is a problem, then the widest possible gate would work for you. However, in my opinion, if your horse isn't listening to you and/or respecting you enough to be able to walk with you through a four foot wide gate, then maybe things aren't so great between you and your horse and you might just have trouble waiting ahead of you regardless of the gate size.

Anyway, I installed a four foot wide entrance gate for me and my horses to walk through into the pastures plus an additional, twelve foot wide tractor gate located near or next to the side of the 4 foot wide man/horse entrance gate. You will want to have gates that can open "in" or "out" depending on how you and/or your horse are feeling you want to enter or exit the pasture on any particular day. Also the gates should firmly click into hardware (latches) on the gate post both to keep the horses in the pasture and to help keep the gates from sagging.

Next you will want to determine for each pasture on which side you want the gate hinges to be placed and on which side you want the gate latch to be placed. Take the time to envision walking your horse to the pasture, holding the lead rope in your hand, getting to the gate with your horse in hand, and then having to open the gate latch and the gate to walk your horse into the pasture. Then ask yourself from which side (left or right) you would want the gate to swing open to determine on which post you will want the gate to be attached, and on which post you will want to have the gate locking hardware (the latch) attached. This is really just personal preference and may not even matter to you. Nevertheless it's something that should be considered, and if it is important to you, should be noted to your fence installer.

Photo #20: Secure Gate Latch Closure Hardware

The next thing to consider is whether or not you want to have **full perimeter fencing**.

When I looked at my pasture fencing on my site plan, I realized that by the time I had fenced in the pastures that I had almost entirely enclosed the property with fencing. Knowing that sooner or later one or more horses would get loose at some point between the pasture and the barn, I decided that it would be smart to pay the relatively small additional cost to enclose the entire property. At first my husband thought I was crazy, and although he agreed to enclose the entire property with fencing, he wasn't sure it was necessary, until ...

Before building the farm, I had asked the owner of the boarding facility where I boarded for years, to stop turning out two of my horses together every day in the same pasture in order to prevent the horses from becoming overly bonded. Horses easily bond. The morning after we arrived at our new farm, my daughter was anxious to do the first turnouts. She innocently proceeded to turn out the two geldings together in one pasture and watched as they happily trotted off together to graze. What none of us realized at that point was that the boarding farm owner had completely ignored my request to stop turning out the horses together all the time, and had, in fact, allowed the horses to form such a tight bond that they were impossible to control if they were turned out together. We realized the situation when my daughter went out to bring in the horses when the two geldings were waiting at the gate. Whereas the one horse was politely waiting to have his halter put over his head (my horse, of course!), his "follower" buddy decided that it was time for "horse brains" to kick in, and, therefore, proceeded to push past my horse and my daughter to dash off in a gallop across the farm. Of course, being bonded pasture buddies, my horse then immediately kicked into his "horse brains" and galloped off across the farm to hook up with his fleeing buddy. It was at this point that I looked over at my husband with that little "I told you so" glance, and he returned with that kind of little look that says reams of unspoken words of validation in return. Right on that first day on the farm, my husband fully understood why I had insisted on full perimeter fencing.

Better to catch your horse safely contained within the fencing of your farm than chasing after him down a busy public road!

NOTE: Whereas 95% of my fencing is flex fence, and all the pasture-enclosing fencing is flex fence, I did use some PVC, four-rail fencing along the property line by the entrance gate that encloses a non-pasture area where I wanted a slightly dressier look to the property, and where horse containment and safety were not a primary issue. Although the PVC fence looks pretty in the limited areas where I have used it, as predicted, rails pop out when deer jump the fencing and don't "quite" clear the fence, and plowed up snow in the winter against the PVC fencing has made it "lean."

There is another way to do fencing that a lot of horse farm owners opt for. It's called **"run-in" paddocks**. These paddocks are just as they sound. The fencing for the paddocks generally opens directly from the barn aisles into a paddock and/or directly from individual horse stalls into paddocks. This is an easy management way to have horses in that you don't have to worry about turning your horses in or out—they turn themselves in or out when they want to. Some people have small, individual paddocks attached to each stall and some have openings into one large, communal paddock/pasture. Other than the obvious issues of having the horses potentially fighting with each other in or over the paddocks, and the potential difficulty of catching a horse that can run off away from you into the paddock/pasture, there is another issue that I feel is the most important reason not to have such paddocks. That issue is being able to have the kind of "control" over your horse that results from having a trusting and respectful relationship built between you and your horse that only results from the constant, positive "handling" of your horse.

Horses are horses. They are not innately born to interact with humans. They have to "learn" to interact with us, and interact with us on our terms. That learning has to be constantly reinforced or the horses will revert to being totally horses. As such, it is important to "handle" your horses every day. Otherwise your horses will be hard to handle or catch, and potentially impossible to handle at important times like when the farrier or the veterinarian needs to work on your horse. Every farrier I have known has complained to me about people who have horses that can't be caught without a lot of trouble and how that causes a lot of wasted time. They also have told me how they refuse to work with horses that are difficult to handle because of potential injury.

Eventually, if you can't control your horses, you will find it difficult to find a farrier or a veterinarian to service your horses. You will also come to find it hard to impossible for you to work with your own horses.

Photo #21: Ian and Emma Peacefully Grazing

So now you've selected your type of fencing, selected a dealer and installer, determined where to install the gates and what sizes and how many, determined whether or not perimeter fencing works for you, and have had the fencing installed. Now you can relax about your pastures. You can also be thrilled with the beauty of your fencing and get excited about how wonderful it will be when your horses are actually peacefully grazing behind all that fencing.

Point of celebration!

Chapter Four

Horse Barn Planning – Builders and Banks

Photo #22: Stage One of Our Barn Construction

Now would be a good time to start firming up **planning the barn** (also the house, if you are concurrently building one, but remember, we won't be talking about that here). You have already roughly planned how large your barn/s and arena will be, but now you have to think beyond just the square footage. It's time to start sketching out some potential designs on plain paper, followed by a better iteration on graph paper, in order to get a feel for the proportions. Yet, as the iterations continue, you might find that no matter how hard you try to get the barn et al looking right, that you still aren't quite sure if it is right or if you have thought of everything. You find that no matter how many iterations you make, that you still don't feel comfortable about taking your drawing out for bids.

What you are feeling is the need for **an architect**.

LESSON #28: Maybe this should be called conundrum number one. Architects don't generally know how to design barns. Barn designers are not necessarily architects. The only certain commonality between the two is that architects and barn designers are both money pits. Worse, whichever one you engage, you will find that whatever you may want, or have drawn, will be criticized, fiercely argued over, and then changed by your designer or architect to their specification in their next set of plans regardless of what you told them to do.

While at a major horse convention and dealer's showcase we were attending in order to gain more information about products before building and buying, my husband became intrigued by a tiny, attractive, earthy voiced, attitude filled, little southern belle. I could see "Miss South" exuding her knowledge and self confidence all over the place as I watched her talking to my husband from across the aisle. After talking to her for a while, my husband excitedly came over to get me from my presumed absorption in the stall front display across from where her booth was located in order to tell me how he had found the perfect solution to our barn planning and drawings dilemma. He had found a "barn designer" who "really knew her stuff."

Being one to never feel too excited about anyone on first pass who is so filled with authority, I reluctantly left my study of stall fronts. My husband guided me over to Miss South's little slot in the long line of booths in the exhibitor's hall. Thankfully the designer was busy expounding to some other potential customers when we came over such that my husband could show me the books of barn designs she had on the table, plus the "finished product" photos she had hanging on her display walls before I would actually have to talk to her.

I didn't really like her as I observed her interacting with the other customers. She was that kind of woman none of us women really like very much. She was just too polished and pushy and had one of those "don't dare contradict me" voices plus attitude. I felt really reluctant to get involved with her especially because I felt sufficiently self confident to be able to design my own barn. Nevertheless, my husband insisted I look through her book of plans where he excitedly pointed out similar designs to the barn we had originally planned to build.

Although I wasn't overwhelmed with anything I was seeing in the drawing book, the photos on the wall were a different story. The finished barns were really attractive and just the kind of look I had in mind for our barn. So, I decided to wait with my husband for our turn to talk to this barn designer.

She was nice enough. She pointed out some interesting things about barns and horses that seemed to have merit. She also had "reasonable" rates (about $1000 for each design iteration) and she was interested in "helping" us with our plans.

Well, "reasonable rates" are only reasonable when they are incurred for one set of plans produced once or perhaps one set of plans produced once, plus one revision. However, rates quickly become unreasonable when each little change means that the entire set of plans will be redrawn again and again, over and over, at $1000 per iteration. Ugh.

As you've guessed, there were many, many iterations and many, many billings for each new set of design drawings. But beyond that annoyingly expensive situation, there was always this nagging little question that I kept playing past my husband that was never really being answered. I kept asking him where the drawings were that showed the actual construction of the barn and ones that would give us a 3D view of "how" the barn would "look" on the inside. All we kept getting were iterations of the interior flat floor plan, and exterior elevation drawings for each side view.

My husband, though thousands of dollars later, finally asked the barn designer to produce the construction drawings so we could take the drawings out for bids from barn builders. Miss South never blinked. She willingly and expediently complied and provided us with drawings and yet another, even bigger, bill. But it was then we found out the awful truth.

Barn designers generally "just design" barns; they don't really know how barns are actually built. That means in the end, we basically had worthless drawings in that they didn't specify anything from a builder's point of view, nor were they necessarily structurally sound. In other words, we could have done these types of drawings ourselves. Maybe our drawings wouldn't have been as "pretty" or "professional" looking as the ones we had just spent so much money on, but they would have had the same impact on the builders bidding on the project.

The good news is that most pole barn dealers do in fact have their own designers who are available to you as a free perk when building your barn with them. These designers supply floor plans and elevation drawings to the pole barn manufacturer who then engineers the actual materials for the barn. The manufacturer will also supply a construction team to build the barn shell from these drawings. What this means is that you can do a rough drawing of your barn plans on your own, and then have the pole barn dealer and manufacturer do the construction drawings that their construction crews will use to build your barn shell (exterior). Again, these construction drawings will be done at no cost to you.

That's the good news about pole barn dealers. The bad news is that you will be put in a "queue" for when your plans will get attention. Also bad is that you will not have any personal contact with their engineers, but, instead, will have to do everything through your local dealer. There also won't be any of the kind of patience, thoroughness, and full disclosure you would get from having your own architect or barn designer plan your barn. In other words, you will quickly find that you are getting what you are paying for. This is a "free" service, and it will feel like a "free" service.

We eventually designed our barn ourselves with floor plan and elevation drawings that we then took to a pole barn dealer.

LESSON #29: What we learned next, and quickly, was that barns are built by barn builders who basically do only the shell plus stalls while they allow you to incorrectly think that their quote includes the whole job. The reality is that you will have to get a house builder (barn finisher) to finish the barn interior at an additional and significant cost. We also learned that many barn builders are basically disreputable and dishonest.

LESSON #30: Amazingly, barn building is generally not a regulated industry. That means in many states, barns and farm buildings do not need building permits, they do not have to "meet code" nor are they necessarily subject to periodic examination by the local building inspector during construction.

In other words you can almost build anything any way you want as a farm building. Plus, barns and farm buildings are not (supposed to be) taxable in the same way that residential buildings are taxable. Tax laws require that property taxes be based on market valuation of your property, and agriculture buildings are not what you would call "best sellers." If you think about barns in regard to the general housing market, most people do not want or need, and certainly aren't willing to pay for, barns. As such, a property with barns, arenas and outbuildings is actually a property with liabilities in the general real estate market. However, the tax assessor will probably go right ahead and assign residential tax assessment figures to your agricultural buildings and tax you accordingly. However, you can and should fight that. A tax assessor might also try and get you to give her your cost figures and tax you on those. Never give the tax assessor your cost figures—or anyone else for that matter—they are none of her business and are illegal as a basis for your assessment. Assessments are only to be based on market value and equivalent properties—never on actual cost!

At first pass, the lack of regulation on agricultural buildings sounds kind of attractive. Don't be fooled. What happens is that the lack of regulation in the industry attracts many "bottom feeders" in the building trades because they can, and will try to, pull anything they can on you because there isn't a building inspector looking over their shoulders (unless you invite one in). I'm not exaggerating. However, do not despair because there are excellent companies that build the basic shell (with roof) for barns and arenas, using their own experienced and competent crews, and eventually you should be able to find an honest barn finisher.

I guess the idea behind the lack of regulation in barn building makes sense overall when you think about the full time farmer who's growing crops and managing animals. When the full time farmer suddenly has a need for a new storage shed, he needs to put something up right away, and cheaply. The same would be true for animal housing. The farmer doesn't have the time or the money to be subjected to building inspectors, codes, taxes, et al. Plus most farmers are probably able to put up the structures they need themselves and don't want anyone "poking around" into their business.

The problem comes in when you are trying to build a farm from scratch and need a "real" builder because it's a large and complicated project. The even bigger problem comes when you try and build a horse farm.

Why?

As soon as you say the word "horse," what the builder hears—and only hears from that moment on in time—is the word "money." Everyone, and most especially those who are planning to work for you, who hears you have horses or sees that you are building a horse farm, will assume you are rich.

The irony is of course that because you have horses, you aren't rich. All your money's gone into your horses! Well, except for those lucky few who actually are rich and have horses.

But try and tell the reality of your financial situation to someone who isn't into horses. I can't tell you how many times I wished I could have said I was building a pig farm, but, oh, well. Just be forewarned that you will be stepping into the world of the sleazy leeches who are experts at bleeding you dry, and actually feel justified in doing so, because they think you are rich and are therefore, somehow, deserving of their exploitation!

Okay. Let's say you have done a barn design and you take your "pretty" drawings out for a bid. Get ready. You are going to find the **bids** to be all over the map and vary by as much as three times and more. There will also be absolutely no rhyme or reason to the prices you will be quoted. The quotes will range from outrageous to bizarre—notice the "reasonable" word is missing here! You will find yourself saying, how could it cost this much? It's basically a big empty building. How could it possibly cost more than the square foot cost of building a house? That's just ridiculous!

Even worse than dealing with most custom wood barn builders is dealing with some of the steel pole barn companies. It's a real "horse race" to determine who can be more disreputable.

So, before we move on, let's take a minute here to decide whether you are planning to build an actual wood barn or a steel-sided, pole barn. Yes, there are other types of barns like canvas, tent like covers, and tube like, corrugated steel structures, but I'm going to assume you want to build something relatively conventional that is permanent and attractive.

Well, in the first place, custom **wood barns** are outrageously expensive. For the price of a four stall wood barn with a hay storage area built above (or in the attic, so to speak), you can get a seven stall, steel sided pole barn, with a wash rack and a grooming stall, an office, bath, tack room, connector aisle AND an arena! Of course, please note that price comparison is totally dependent upon which steel pole barn dealer you select.

In our experience, the "leader" in the pole barn field, or at least the one who does the most advertising, is outrageously deceptive in their pricing, and is way overpriced in the end. In fact, a price my husband received was so incredible to me for "just" an empty, steel sided structure that I argued unpleasantly with my husband that he surely HAD to be wrong! I was positive that the quote he had been given was for a finished barn. I was so sure that I was right, and that my husband was wrong, that I made him schedule another meeting to review the quote with the supplier. Well, sure enough, I was flat out wrong! The outrageous quote was actually, in fact, for an empty, unfinished barn! But finding out the truth of that quote took an hour long meeting with the dealer and a lot of intensive questioning before the reality was fully revealed by the dealer .

LESSON #31: When you get a quote for a steel sided, pole barn, it is only for the shell which includes either a basic steel roof or a roof "ready" for shingles, their stock exterior windows and doors, and exterior wall, steel siding. Nothing else is included except maybe basic, cheap stalls.

NOTE: The structure would have a dirt floor in that floors are also "not included" in the basic quote. Amazing.

In our experience, the "leader" in **steel pole barns** will happily allow you to believe that you are getting a finished horse barn with their outrageous quote, and show you lots of photos of finished barns they have "done." But upon closer questioning, you will find out that most likely all you are getting in your quote is the basic pole structure, covered with steel, and with the cheapest possible windows and doors and roof. Typically in the quote, you will also be getting the minimum size structural wood that barely holds up the load of the building which includes their estimate for a snow load on the roof (if appropriate for your area), low cost framing on doors and windows, a plain steel roof (not a fancy steel roof like you see on banks and upscale shopping centers), or shingle-ready roof (particle board without shingles), and a dirt floor. This type of quote may be okay for a hay barn where there is minimal interior finishing required, but it is grossly incomplete and misleading for a horse barn.

To help you keep your sanity, you should know that the basic steel pole barn or arena **shell** with a finished roof and a minimal number of doors and windows and a dirt floor should cost from $10 to $20 per square foot erected on your leveled site.

LESSON #32: Be sure to ask for (and get) complete specifications of everything that is included in the wood barn or pole barn quote.

I built a steel sided, pole barn and, therefore, I can't give you much information about wood barn builders. However, I did go to two wood barn builders to get quotes. The first builder did quote on an entire, finished barn—both exterior and interior. The second builder quoted only on the basic shell without a finished interior. Both quotes were so outrageous that we didn't pursue any further discussion with them. As such, the following information that I offer will assume that you are building a steel sided, pole barn.

Because you will need to have **colors** selected for your pole barn, this is a good time to start thinking about what colors you want. (Also you should make a note to ask your pole barn dealer how many years the "colors" are guaranteed not to fade.) It's best to have two basic colors for your barn—one for the main color of the siding, and then one for the trim, including the all-important **cross bucks** on the barn doors. Cross bucks make a HUGE difference in the aesthetics of your barn in that cross bucks are another one of those barn icon traditions. However, having cross bucks does add some cost. You should also have wainscoting around the bottom of the barn and arena exterior walls in your contrasting, secondary color which will make a HUGE difference in your barn's appearance. Remember that white siding can be easily stained by rust in water and that rusty water is very common. Therefore, it's advisable to keep any white surfaces a few feet above ground level. In other words, it would be a mistake to have an entirely white barn without any wainscoting of a darker color around the bottom exterior or to choose white for your **wainscoting** color.

Selecting your barn colors is important. It will determine daily satisfaction over the appearance of your barn, and it will be the basis for all products you select to coordinate with your barn such as custom horse blankets, coolers, jackets, etc. Drive around and look at the colors on various steel pole barn structures to help you decide on colors. There are also interactive sites on the pole barn websites that allow you to "play" with different color schemes on a barn icon. This will help you do a final call on colors including your roofing color choice.

And as long as you're thinking about colors, this is also a good time to start thinking of a **name for your farm** as well—which is not as easy a task as it seems it should be. But once you do come up with a name for your farm, think about an image you would like to have that represents your farm name and/or purpose. You can go online and find a lot of ideas under "wood sign maker" websites to get the creative side of your mind thinking. However, if you're creative enough on your own, you can just take your design directly to a sign maker or even deal with one remotely based on the supplier's willingness to do so from his website.

If you want a really unique and professional edge to your logo and farm sign, then you might want to consider taking your farm name and ideas to a graphic artist and hiring him or her to come up with some designs for you that incorporate a **logo**, farm name, and farm colors. The cost for the custom, graphic design will range between about $500 and $1K. Once you've decided on a design, the graphic artist will supply you with a CD that contains all the formatting required for various suppliers to use your logo, including everyone from sign makers, to jacket makers, to blanket embroiderers.

Photo #23: Wood Carved Farm Sign with Copyrighted Farm Icon Logo

Drive around your area and start looking at wood **signs** to also give you an idea of how big of a sign you want (yes, you should actually take a tape measure or yard stick along with you and jump out of your car and measure signs that seem to be the right size for you), note the shape of the sign, as well as how tall and what kind of support poles you want with your sign. If you find a good looking sign, go in and ask the business who made the sign for them. This is the best way to find a good, local sign maker. Once you have determined the size and shape of your sign, and the height and type of poles for your sign, it's never too early to go ahead and contact the **sign maker** and have the sign made. Our wood carved farm sign with posts cost us about $1500. In fact, it's really exciting the day your sign arrives and you can install it and feel like the farm is really yours—regardless of the muddy stage of construction that is currently going on behind the sign! But do protect your sign from being knocked over by placing BIG boulders around its base or someone WILL knock it over for you.

I shouldn't fail to note that there are cheaper ways to have a farm sign. The farm where I boarded for many years had an attractive sign "carved" out of foam. I don't have any personal experience with such suppliers, but the sign was nice and stood the test of time. I'd imagine it was relatively cheap too in that the farm owners were cheap about everything else so why not about their sign? There are also, of course, many degrees of plastic signs that can be produced to display your farm name. However, for me, I just love wood signs so much—when they are done correctly—that I was willing to spend the extra money to have one made and have found that I enjoy the "fruit" of that investment every time I enter the farm, as I'm sure you will as well.

LESSON #33: You will require a second barn builder to finish your barn other than the pole barn/arena builder who "built" your steel pole barn and arena shells.

You can elect to have the pole barn people partially finish off your barn interior, but you will get neither value nor quality for your money. You will also be unable to finish off the barn with any extensive personalization. However, what you do need to specify with your first builder--your pole barn builder--is any exterior metal shell modifications you will be making such as custom, upgraded windows and doors from other suppliers. Any customization that affects the basic shell is handled by the pole barn builder. The rest of the project and customized modifications can be handled by the second barn builder, or barn builder finisher. We specified custom stall windows and a custom front door for our barn as well as custom stall fronts and special water proof, energy efficient, long life, barn lighting fixtures.

For the best value and satisfaction, you should hire a second builder who will be the one who "finishes" off the interior of your barn and arena and customizes it for you. The second builder constructs the wood, tongue and groove stall walls, installs the stall fronts you will specify, subcontracts the concrete floors, subcontracts all the other trades such as HVAC, electrical, plumbing and drywall, installs the electrical fixtures you will specify in the barn and the arena, installs any additional wood overlaying the steel siding in the aisles, finishes the wash rack, the tack room, the office, the bathroom, the arena kick wall, and on and on. This type of builder is best selected from among the remodeling type builders. He is the one who knows how to do carpentry work and can orchestrate the plumber, electrician, etc.

So, again, as with fencing, there is more than one phase or supplier with pole barn structures. The cost of the steel structure is just Phase One. You can figure the interior finishing of the barn as Phase Two and that the finishing of the interior will approximately double the cost of the original barn steel structure quote by the time you are done. However, for a complete barn, anything more than double the barn shell price is out of line. What we found was that most barn builders would take the cost of the steel pole building structure and multiply by three or four, and then give us that amount for their "completed barn" quote. Don't forget that the arena is basically just an empty shell and its complete cost would not come anywhere close to the cost of completing the horse barn.

LESSON #34: Do NOT hire the same builder to build your house and your barn. Hire a third type of builder for your house - a house builder.

You might think that lesson sounds silly in a way. It might seem kind of obvious to you once I've said it such that you're probably thinking I mean that your house will end up looking like a barn or vice versa. But, no, that's not the actual "problem" with such a linked building situation.

First of all, as I previously mentioned, house builders have NO clue as to how to build a barn and they have LITTLE interest in learning. They also have a tendency to care little to nothing about large animals, and feel that any kind of minimal effort on their part to build an animal housing structure is just fine regardless of how inappropriate their work is. Plus they will NOT listen to what you want done for the comfort/benefit of your horses or "hear" anything you have said. They just do NOT connect.

Second, most barn builders, being the bottom feeders that they are, if given an opportunity to link the building of the barn to the building of your house, will take full advantage of "stalling" the completion of the barn, use up all your construction loan on the house, and then rape you for money to finish building the barn. Yes, they are that scummy.

In general, you can pretty much plan on firing your first barn finisher builder and expect him to leave town with a big chunk of your money. I haven't met anyone who has had a different experience. This is true whether or not the house and barn projects are linked or not linked.

Oh, come to think of it, now would be a good time to talk about money. After all, you will most likely have to finance the project. Herein lays room for a lot of surprise and disappointment.

LESSON #35: You will have to take out the <u>construction loan</u> for the entire job yourself.

As I briefly mentioned earlier, a bank will generally not give a construction loan for just a non-commercial barn and outbuildings. If you need a loan, you will have to get a house construction loan big enough to cover the costs of both the house and the barn/arena/outbuildings. Or you can build a commercial facility and try to get a small business loan, if you have enough collateral. Or, of course, you can pay cash!

When you take out the construction loan, you will quickly find that the bank is not your best friend. First, if you don't know better than to request an **"owner's" construction loan,** the bank will give you a construction loan document to sign that is meant for builders, and then manage the distribution of the money as if your subcontractors were the builders. They will collect huge amounts of interest money on your loan from you, yet they will not appropriately monitor the signing off of draw requests by your builder to your advantage.

When you have a "builder's" construction loan, the bank is the one with the dominant authority to decide when to release the money on the draw requests—not you. Then because the bank doesn't care if your builder is being honest in his draw against your loan, they will be happy to hand over your money to your builder as if he had taken out the loan, whether or not he is lying and cheating on his draw request sheets, as long as you are paying the interest to the bank on the construction loan.

With an "owner's construction loan," you can specify that only you have the authority—not the bank—to review the draw requests before the bank releases any of the construction loan money to the builder.

Even if you go in and threaten the bank to stop giving your builder your money before you go over the draw requests and get verification for costs, they will likely smile, give you lip service, and still hand over your money to the builder if you have a "builder's" construction loan. And, then, of course, if you don't pay the interest on the loan, they will be happy to immediately call in the full amount of the loan from you!

So, you must be sure to have an "owner's" construction loan and explicitly specify that only you have the exclusive right to release the money to the builder when YOU are satisfied with the work he has done and that all liens have been removed. You should also put in the loan contract with the bank that if they release your money, then they will incur financial penalties. You must also be sure to define the penalties in the contract.

Even **liens** that can and will be put up against your property are written to the advantage of the supplier and meant to disadvantage you the home/barn owner.

The subcontractors must give you written lien waivers saying they have been paid in full for the work they have done up to that point in time. The written lien waivers must be handed to you before you hand them the checks to pay for their services. It is a two step process. First, the subcontractor signs the lien waiver that says if you pay a certain amount of money, that he will wave his rights to place any liens against your property.

Then, second, when the subcontractor gets his money, he must sign the document that says he has received full payment.

So if the builder is given free run to get your money from the bank, he will not necessarily care if he bothers to get the signed lien waivers from the subcontractors as long as he gets them to go away while he's building. Without signed lien waivers, the subcontractors can come back at you, at or after closing, and demand payment on already paid bills or demand more money because you have no proof to defend against the demand for money.

One of the fundamental problems in building a farm as an individual (not a builder/developer) is that you will have to outright purchase the land for your farm with cash so that you can get a construction loan. The land is your equity for the loan. Otherwise, if you didn't own the land you were building on, then it would be like the bank was giving you money to build on someone else's property. The bank won't do that and risk losing everything. At the same time, because you own the land, any builder you hire has no collateral on which he can base a construction loan for your project with the bank—because you own the land and he does not. Therefore, you necessarily become the developer of the land and also the one who has to take out the construction loan as the owner and general contractor. But, then, because the builder "has nothing to lose" as such--because he has no financial investment in your project--you become "at the mercy" of your builder who can create endless nightmares such as schedule delays for you should he be dishonest.

LESSON #36: Do everything you can to check out your builders credentials and ethics—both for the barn and the house.

Get several **references** from your builders. One of the most effective ways to find out who's good and who's bad in the area is to drive up to new barns you pass and ask the owner who built their barn and how it was to deal with that builder—both the metal pole barn supplier/builder and the barn finisher builder. Most people will be openly honest about the whole situation and more than happy to share their experiences as cathartic pain relief! However, sometimes it's hard on your part to accept all the negativity you might hear from the barn owners regarding their various builders, and you will find that your mind starts to discount the information it's receiving as excessively bitter.

My advice would be to keep an open mind and listen carefully to everything said. Also of course you can directly ask potential builders for references. But then, you must be sure and call those given references and talk to the people about their satisfaction with the work, and what it was like dealing with, the builder. Then ask the references if you can go see their house and/or barn—and then go see it! Don't feel that it is enough to have your builder give you references or that because he had references to give, that they must be valid and happy customers or non-relatives. Check it out.

Also ask your builders for their builder **state license numbers** and then go online to your state's website, type in the builder's license number, and check to see (a) if he has a valid license and it is current; or (b) if his license has been revoked; and (c) if any complaints have been registered against the builder.

Know who you are dealing with. Otherwise, as an example…

We had lined up a barn builder (finisher) and had made the mistake of making him our house builder as well in that he had supposedly built both types of structures. He was the only builder who gave us quotes that seemed to be in line with what both barns and houses should cost. (We later found that his original quotes were just a sham to get us to contract with him.) We had also taken out the construction loan with the "first" bank. We had made it clear to the builder that we wanted the barn to go up first and be finished as soon as possible so that we could move our horses into the barn and get out of where we were boarding. (Remember that our township allowed you to have a barn with animals on the property without you having to live on the property as well.) We had a local condo to live in while the house was being built and wanted to vacate our house on the other side of the state and sell that house as soon as we could move our horses into our own barn. As previously noted, linking the building of the house and the barn with one builder was the first BIG mistake we made that haunted us to the very end of the project in that the builder continued to delay finishing the barn while running up the costs of finishing the house. This also prevented us from moving our horses to our own barn and selling our other house. The end result was a huge loss of money between the continued cost of boarding the horses, and the fact that the real estate market happened to collapse during the two year construction delay. We wound up having to sell our house for hundreds of thousands of dollars less than it was worth when it should have first been listed for sale and sold.

But then there was the problem with the bank and the construction loan.

We went to sign the papers for the construction loan at a title company which is the custom in our state. We took the time to look over the documents as the loan agent quickly reviewed with us what each page "meant." When we got about mid way through the package of loan papers, my husband noted that there were some documents missing that he had been told would be included and therefore he said that we could not continue with the closing. The loan agent was quick to pressure us saying there weren't any missing papers and that we should just keep signing the papers. She also "threatened" us that if we didn't sign all the papers by the end of the day that the loan would be withdrawn.

Of course, by the time you get to the point of signing the papers for the construction loan, you have put in a lot of time, money and emotion into that moment, and therefore you are ordinarily inclined to fold under such "threatening" pressure by the agent to sign, and sign now, in order to keep the project moving forward.

However, we didn't fold to her pressure and said that we needed to call the bank loan officer and ask where the missing papers were, and to produce them, before we could finish signing the documents. So, much to the agent's dismay, we got up and left the conference room and immediately put in a call to the bank loan officer who was of course unavailable on the first pass. Eventually, however, the loan officer called us back and assured us that all the documents that needed to be signed were there and that we should proceed with the closing. We had no reason not to take him at his word and so we returned to the title company to sign the documents. Upon signing, we were told that the loan had been activated and that draw requests could now be made against the loan.

Great!

The first draw on the construction loan was for $50,000 which the builder immediately made. The $50,000 was supposed to be used to order the barn materials and begin the barn construction. The use of the first draw had been clearly communicated by us to the builder and he acknowledged he fully understood. The loan was granted in October and the builder had promised the barn would be finished for us by December or for sure by January at the latest. We were excited of course and everything looked like it would work out just right. We could move the horses into the barn by January and put our house on the market and move into our new house on the farm in May. My husband was in immediate contact with the builder after the release of the $50,000 who told my husband that he had used the money to order materials and that building would soon be underway.

However, when we drove across the state to check on the construction progress a couple of weeks after the first $50,000 draw, we didn't find anything under way as far as a barn and arena were concerned. What we found instead was that the foundation had been dug for the house and the footings for the house had been put in—incorrectly, I should note. You can imagine how very upset, disappointed, and confused we were.

NOTE: Make a friend of the **building inspector**. He can be a tremendous help in double checking that your builder is doing the work correctly on your project and make your builder correct anything that he has done wrong.

In an effort to avoid alienating the builder from day one of the project, we decided to wait another week to see if the builder would start anything on the barn. But the week went by and nothing happened. Now we felt really concerned about what the builder had done with the $50,000 draw he had taken upon the granting of the loan and was supposed to use for starting the barn. So, we called the bank's accounting firm and asked for a cost accounting. When we called for an accounting, the bank's accounting firm informed us that there had not been any cost accounting done at all against the $50,000 draw, that the money had simply been released to the builder in a lump sum, and that further, the bank had issued a "stop payment" on the rest of the loan.

We were shocked. Why wasn't there an accounting? Worse, why had a stop payment been issued on our loan? Why hadn't the builder told us any of this? Why hadn't the bank told us any of this? What in the world was going on?

Then, in early November, I got a phone call from the bank. The loan officer on the other end of the line, who was the one who had approved the loan in the first place, now told me he was "overnighting" a couple of pages that had been missing from the loan package of documents when we signed it, and that they needed to be signed immediately and returned to him via overnight mail.

"Missing pages!" I thought. "That sure sounds familiar!" I immediately saw a red flag. When my husband got home I told him about the call and said that something was obviously, and very suspiciously, wrong.

When we received the documents from the bank the next day we were shocked.

The documents stated that the bank would not take the responsibility for making our builder adhere to the cost accounting sheets agreed to before the loan was granted nor would they monitor the builder's draw requests. In addition, the bank wanted us to approve that our current house would be part of the loan guarantee such that if the builder did something that violated the terms and conditions of the loan, that the bank could take our house from us and immediately sell it out from under us, at an artificially depressed price established by their appraiser based on a "drive by" appraisal on the exterior of our home. The bank further told us that they decided they didn't like our builder (even though they had told us previously that they had vetted the builder and approved him as our builder before issuing us the loan which had given us the confidence to go ahead and use him as our builder), but they wouldn't tell us why. They said their decision was based on "something" about our builder, and they were basically forcing us to sign the papers by stating that if we didn't agree to sign the "new" (or previously missing) loan papers within 48 hours, that the bank would immediately pull the loan and we would owe them the $50,000 the builder had already spent, payable immediately, or they would sue us for the money.

If the bank did in fact sue us, then any and all construction on the property would stop until the lawsuit was settled, and that settlement could take years.

We said, "They can't do that!"

But the bank claimed that we had signed a document in that large pile of papers we signed on the construction loan "signing day" that had some "hidden" loose wording that applied to this situation and supposedly allowed the bank to change the conditions of the loan at any time and in any way they wanted.

Oh, that's why the bank hadn't cared about missing pages on the day we signed the construction loan! They knew they had recourse at any time to do anything they wanted, and we didn't!

The reality of the situation was that what the bank was trying to do to us was illegal. No one can retroactively and unilaterally modify a contract because to do so is to create a new contract thereby voiding the previous one. But practically, it didn't matter because we couldn't take the time, nor did we want to spend the money, to sue the bank. By the way, this bank was one of the largest national banks.

I don't know if you have had any dealings with trying to argue contracts, with attorneys in general or in the court system, but there are two overall lessons we have learned over the years. First is that lawyers will tell you little, do little for you, talk out of both sides of their mouths, string you out endlessly in order to run up the bill, and then, in the end, regardless of what they told you in the beginning, tell you now that you don't have a case and/or that you can't "win" in court. Second is that people and corporations lie freely, will do anything they want regardless of any document you or they have signed, will be happy to run up an attorney's fee for you by finding new and continual ways to force you to defend yourself, won't take you seriously until you take them to court (which generally speaking will cost you more than you'll ever recover), and will most likely find some moot point to win against you or resort to "buying" the court to rule in their favor.

So we knew we weren't going to "win" against the bank's little game they were pulling on us. Meanwhile, the panic was that they were going to pull the loan and we'd immediately owe them $50,000—plus we wouldn't have a construction loan, construction would halt for an indefinite period, and we weren't sure what the bank had found out was "wrong" with our builder which could potentially keep any bank from granting us a construction loan. Ugh. Oh, and of course, our builder had the $50,000 which meant that our cash was in his hands and basically unrecoverable.

<u>**NOTE:**</u> Be suspicious of a builder or subcontractor who asks you to buy the materials for him with your construction loan money. The loan is meant to pay for progress in construction when materials are combined with labor and something is built. Such a request is a red flag that the builder or subcontractor does NOT have a credit line. The builder should be using his line of credit to buy materials. There is some negative

reason why a builder or subcontractor doesn't have a credit line. I would recommend you ask your candidate for a builder upfront what the size of his credit line is, and then have it written into your contract with the builder that he must use his credit line to buy materials.

Well, we did manage to get another bank to give us a loan within 48 hours and pay the first bank the $50,000 which saved us, but frankly it was only because we had "friends" who had friends within the banking world who were able to help us expedite things.

So you can see that banks can be bad in all directions—both by potentially disguising the terms and conditions in the loan, or being able to arbitrarily pull your loan, and then in the way they administer the draw requests on your loan.

LESSON #37: Deal with a reputable bank that operates with a local loan officer face-to-face with you and have your attorney review and approve the contract for the construction loan before you sign it—making sure that it is an "owner's construction loan"—NOT a "builder's construction loan." Also meet with the loan officer at the bank before you sign anything, thoroughly discuss and reach an understanding on how your money will be granted to, and accounted for by, the builder, and get that all in writing for your attorney to review and approve.

Take the time to check out the bank. The first bank we dealt with was a huge, national bank that managed their construction loans remotely through a service firm. It was much later that we heard that this bank, and several other of the big, national banks, had in fact been engaging in unethical behaviors and were being investigated. So, once again, bigger isn't always better! Also, although the bank we dealt with in the end was a local, reputable bank, because it was a commerce bank, it was still more builder friendly than was appropriate for our situation.

LESSON #38: You will have to do constant (preferably daily) "checking" as to the progress on the construction site, and constant accounting, which means a minimum of weekly, accounting spread sheet updates and discussions with all builders.

As mentioned, you should be the one and only who reviews and then signs off on any **draw request** on your loan before the bank releases any money to your builder. Otherwise you will have no idea where your money is going or how much of your money has been spent. Without personally and carefully controlling the money, you could potentially find that the loan may not cover the cost of the construction which would be a disaster. This accounting and checking up on the builder and the construction is very time consuming.

So, let's review here. Your pastures are in and your fences are up. You have a barn design, know what type of construction you want, have found an honest barn shell supplier/builder, and you have found both your second, interior barn finisher/builder, and your house builder. You've gotten a construction loan and have a bank approved spreadsheet accounting system that has you firmly in control of your money. Basically you are on your way.

Chapter Five

Hay Barn Planning, Commercial Facilities, Waste Disposal

Oh, but wait a minute! What time of year will your horse barn be done? Will it be done just in time for the first cutting of hay? Or will it be ready in the dead of winter when there potentially isn't any hay available for sale? Or maybe not enough hay? Or maybe hay at a ridiculously high premium meant to exploit those who haven't planned ahead.

LESSON #39: Unless you are absolutely sure that your horse barn will be completed at the perfect time for you to get a year's worth of first and second cutting hay for your horses, you will have to immediately put up a hay barn and get in as much hay as possible now.

First of all I am assuming here that you will be buying a year's worth of hay to store, and that you will be building a separate structure for your hay storage. Please do that—have a separate hay building structure. Although I tried to plan "cheaper" ways of storing hay by incorporating the hay storage in with a main barn in some of the barn design iterations I went through, there just wasn't any way to safely store hay in the same building in which the horses were housed. Not only is hay a huge fire hazard, but storage above the barn aisles cuts off proper air circulation in the barn for your horses plus makes a lot of dust and debris droppings into the air of the barn; both of which are not good for your horse's lungs and general good health.

Photo #24: Hay Barn Early Construction -- "Poles"

The building cycle for the erection of the horse barn/arena shell and the completion of the interior should only be a total of about 6 to 8 months. Lots of things can go wrong that will delay the completion of your barn/arena. It's just not a good idea to count on any precise completion date. The best you can do is to have a rough estimate of when your barn/arena will be done and you can move in your horses.

Photo #25: Hay Barn Construction

Unless your horse barn/arena and hay barn will be completed exactly when the new season of hay is being cut so that you can buy and store a year's worth of hay or you don't have to move your horses into your barn until that time, you will need to put up the hay barn independent of the rest of the horse barn/arena construction. You will have to pay cash for this hay barn project outside the construction loan in order to get the pole barn people to construct the hay barn ahead of the rest of the project.

The hay barn needs to be completed before the hay cutting season prior to the completion of your horse barn/arena. In this way, you can calculate, buy, and store the amount of hay you will need to fill the gap in time between when your horse barn/arena construction is finished and your horses have moved into your barn, and the next hay cutting season.

For example: Hay cutting begins in late June in our area. So here, your hay barn must be built by June 1st in the year XXXX. Then you would estimate when your horse barn would be completed. Let's say you estimate the horse barn to be ready for your horses in December of XXXX. So you would calculate how many bales of hay you will need for your horses between December of XXXX and the next June of year XXXY when

Photo #26: Hay Barn Construction Progresses

Photo #27: Hay Barn Completed -- Note Crossbucks on Slider Doors

the new hay cutting season begins. You would then purchase that number of bales of hay in June of year XXXX, so that you will have hay for your horses when they move into your new barn in December of XXXX that will last until the next cutting season in June of XXXY. And, it's always best to overestimate the amount of hay you will need, rather than to run short.

The hay barn doesn't have to be fancy as such and therefore doesn't require much design time. The most important thing about a hay barn is calculating how big you want it to be. You must also decide if your hay barn will only be used for hay storage or if it will also be storing your farm vehicles and equipment, like your tractor and Gator. Because you will want your hay barn to be consistent in its exterior design with your main barn and arena, hopefully you will have a good feel of colors and styles for your overall project so that you can have the hay barn match the rest of the project—and don't forget about cross bucks!

Photo #28: Farm Equipment in Hay Barn, Fluorescent Lights on Trusses, and Hay Bales Stacked on Pallets

LESSON #40: If you can afford to have a separate barn for your farm equipment or a shed roof storage area off your arena roof (made simply by extending the arena roof to make an "open," covered, porch-like storage area) it is better than storing the equipment in the hay barn.

First of all, you will need to get your tractor and Gator, etc., in and out of the storage area all of the time. Opening and closing the hay barn doors and/or juggling the parking of the equipment within the hay barn, can become a real drag. But even more important, we found that continually bringing the farm equipment in and out of the barn brings too much moisture into the hay barn, especially in the winter when the tires get jammed with snow that melts onto the barn floor. And the problem with moisture is that it breeds mold!

My hay barn is 36 feet by 36 feet square with 16 foot high exterior walls. When combined with the 6:12 roof pitch, the side walls make a floor to center ceiling peak height within the hay barn of 25 feet. Because of the roof trusses, the height for hay storage is only 16 feet. But we were able to store 900 bales of hay along three walls of the barn on pallets (enough for 3 horses for a year) that extended out about 12 feet on one side and 8 feet on two other sides, leaving the center of the barn open with enough room for storage of a John Deere 3000 series compact utility tractor with a front bucket attached and a rear attachment like a rotary cutter or a box blade, a Gator HPX 4x4, other power equipment, and a workbench. From this size hay barn, you can estimate the size you will need for your hay barn according to your hay storage need. Remember that you can't stack the hay barn wall to wall with hay bales. The middle of the hay barn has to be open for you to access the different types of hay in the barn, and for the hay farmer to be able to back into the barn to deliver the hay.

Here's the floor plan for our hay barn.

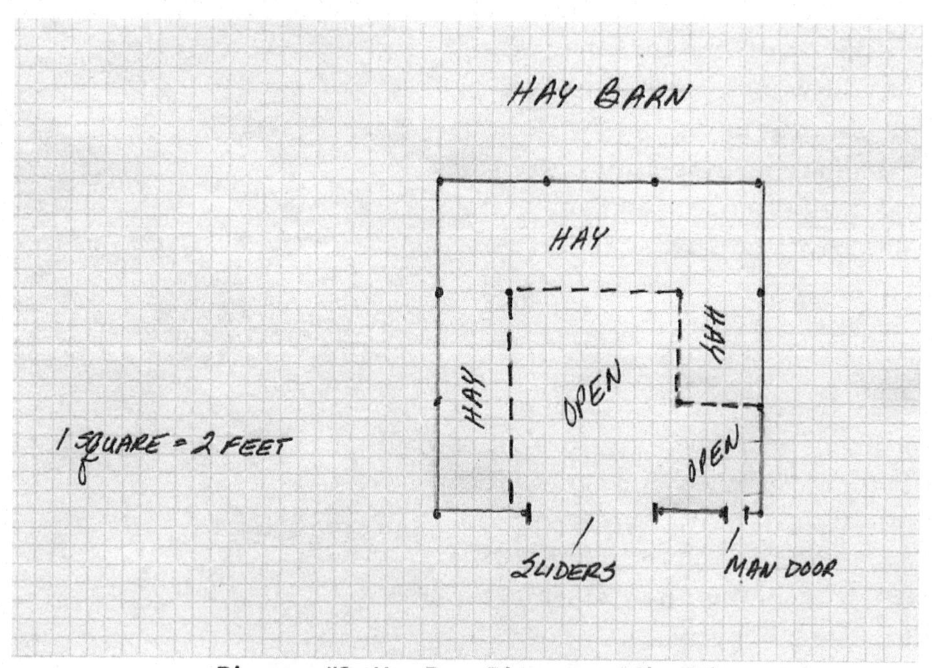

Diagram #2: Hay Barn Diagram – 36' x 36'

NOTE: The roof pitch on the hay barn should be nice and high—with a minimum 6:12 pitch. Ideally for appearance, the roof pitch on your hay barn should be the same that you will have for your horse barn. The arena will have a lesser pitch - in our case it is 3.5: 12. We were told this was the maximum pitch for trusses that span 72 feet to be able to be transported on a truck so they fit under Interstate highway bridges.

LESSON #41: The slider entry doors into the hay barn must have a minimum door height of 11 feet and a minimum door width of 15 feet (two doors each 7′ 6″ wide) in order for delivery vehicles to be able to enter the hay barn.

NOTE: Don't forget to have at least one **man door** into your hay barn! Remember that your hay barn doors will most likely be just relatively thin sheets of metal, and, therefore, will always have to be locked into place when not in use or they will blow in the wind and potentially get damaged or blow off of the hay barn. As such, you will need to be able to enter your hay barn through a locking man door—with an exterior light fixture above, or next to, the man door.

Calculating how many bales of hay you will want for your horses will depend on varying factors and can become more of an "art" than a science to determine. First you will need to consider how many horses you will be feeding. Next you will need to determine whether or not you will be turning out your horses to graze for most of the day. That calculation can be tricky at this point because you don't know now how your horses will react to their new pastures and how much time they will actually spend in them. Then you will have to know if your horses are "big" eaters or not. You will also have to know if your horses "waste" a lot of hay which is tricky to know at this point, because "wasting" hay has as much to do with the quality of the hay you will now be getting from new sources as it does with how finicky the horses actually are.

But overall, a general calculation is to estimate that an average horse of about 1000 lbs will eat at least a half bale of hay a day in the warm weather if they are not turned out for many hours to graze, and at least one bale of hay per day in the winter. Personally, I calculate a little less than a bale a day per horse for a full year (12 months). That means about 300 bales per horse per year which roughly compensates for having enough hay to feed considering waste or other unforeseen issues.

NOTE: Don't panic if you run out of hay, especially the first year when your calculations may have been slightly off the mark. There are usually area wide "hay auctions" held where you can buy some hay (but shouldn't be relied upon for your basic hay supply or the bulk of your hay supply), as well as hay farmers who have local "left over" supplies available for sale. It's better to have your regular hay farmer, who is your supplier, represent you in any off season hay acquisitions. He will know what local suppliers he can call first to determine if they have hay available. He can also check the quality of the hay for you and if it has been correctly stored. The problem with having to go to hay auctions is that there are deceitful people who "hide" bad bales of hay in the center of the hay sale cart which you do not uncover until you are back unloading the hay at your farm. Also, you will pay a premium for the off season hay and hence the "auction" word!

<u>LESSON #42:</u> Cement the floor of the hay barn no matter what anyone tells you. If you don't cement the floor, you will have trouble with mold in your hay barn, and mold is deadly for horses if it gets into the hay bales and then into the horses' nostrils and lungs. You will still need to elevate the hay bales on pallets to get air circulation under the bales to prevent mold.

<u>NOTE:</u> If you don't cement the floor, then you must cover the dirt floor with heavy duty, thick "visqueen" plastic sheets and then lift the hay bales off the floor by putting them on wood pallets to allow for air flow under the hay bale stack in order to prevent moisture build up and subsequent mold.

The problem with **<u>mold</u>** occurs when the horse grips hold of a moldy hay flake and rips it apart to eat. Clouds of mold spores shoot out into the air and go right up the horse's nostrils and into his lungs. The mold spores are deadly to horses. One big shot of the mold spores and the horse's lungs are permanently destroyed. So being vigilant about checking for mold in your hay is mandatory, and feeding moldy hay must be avoided at all costs so as not to risk your horse's life.

However, you can resell your moldy hay to cow farmers in that cows have no problem eating moldy hay. Or if you don't want to go looking for cow farmers, you can negotiate with your hay farmer to remove the moldy hay from your barn at no cost to you by allowing him to resell the moldy hay for his profit to the cow farmer. Once the moldy hay is removed from your barn, you must make a point of thoroughly cleaning out your hay barn to try and minimize the presence of mold spores.

As a side note, I had always wondered when I took a riding clinic in England, why we had to soak our hay bales in a garbage can of water overnight before we fed the hay to the horses in the morning. It seemed like such a cumbersome, messy task. (Picking up a hay bale is heavy enough—try picking up a hay bale saturated with water! It's incredibly heavy!) Being one of the two "ugly Americans" in the clinic, however, I wasn't about to ask anything about why we were soaking the hay and be more "on the outs" than I already was just for being an American. Then, coincidentally, our hay farmer here mentioned to us, while removing our moldy hay for resale, how one of his customers had commented on how in England, where she was from, that ALL hay was moldy due to the dampness of the weather. As such, they had to soak all their hay in water before they fed it to the horses so as to wet down the mold in the hay and avoid contaminating the breathing passages and lungs of the horses. There was my answer! Yes, it was an answer received about 16 years after the fact, but at least I had an answer!

<u>LESSON #43:</u> Put <u>downspouts</u> on the hay barn (as well as all barns and the house), plus long downspout extenders at the bottom of the downspouts that direct the roof

rain water way out and away from the barn. Do not use an underground drainage system for the downspouts.

The first sign that you haven't extended the gutter downspouts out far enough will be little ponds of standing water at the bottom of the downspout. In a heavy rain, those little ponds will grow into lakes that back up and put unwanted moisture into your barns and arena.

And then there's our idiot first builder who decided to put underground drains from the three downspouts on front of the barn that all fed into one central drainage pipe that was supposed to carry all the water away from the barn.

That sounded right. In fact, it sounded like a good idea, right? However, ...

It was when we were having one of those three day rain sieges that are relatively common here in the fall of the year. It had rained enough by day two, that I was beginning to feel a little nervous about the level of our drainage pond water. But I wasn't particularly worried about anything else in that we had all our gutters and downspouts in place on the house and all the out buildings.

On the morning of day three of the rain, I entered the barn to grain and hay the horses blowing into the aisle with the gust of wind that entered with me through the barn door. I was a little caught up in myself being as drenched and wind blown as I was. Nevertheless, it caught my eye that Ian, who is in the first stall by the barn door entrance, was standing rather awkwardly in the back of his stall, almost as if he had been stuffed back there rather than stepping forward as usual to greet me. His overall demeanor was strange as well, although I couldn't exactly read its meaning as I hurried to throw some hay into all the stalls. But as busy and focused as I was on my mission, it instinctively seemed to me that Ian was concerned about something and trying to give me a "heads up" as he stood there kind of scrunched into place.

I continued on my mission to throw hay into the stalls. Along my patterned path, I noted that Emma, in the middle stall, was also standing rather oddly at the back of her stall in a similar kind of way to Ian. Nevertheless, Alex, in the third stall, didn't seem to be particularly suffering from the same "problem" the other two horses were experiencing. So, the behavior of Emma and Ian didn't take a prominent position in my mind at the moment as I continued to complete my tasks.

Then after haying, I began the graining process. Scoop in hand, I opened Ian's stall to put his grain in his bowl. I opened his stall door with the same rigidly patterned behavior I had on any other day, expecting everything to be as it always was. After all, why wouldn't everything be as it always was? Exactly. Why wouldn't it be?

However, when I stepped into Ian's stall, I found out that today was not just like every other day, because when I stepped into his stall, to my total unexpected horror, I dropped down about a foot!

That certainly got my attention! My eyes now immediately redirected themselves and riveted on the stall floor to find that the stall floor had indeed collapsed down about a foot from the mid point of the stall, all the way up to the front of the stall. Yes, this was the situation my eyes were seeing, but, at the same time, this was a situation my eyes were finding very hard to believe they were seeing!

What in the world had happened? Dear Lord, it seemed that the stall floors had actually somehow collapsed! The horses weren't randomly behaving strangely, but were actually "stuffed" into the back of their stalls by lack of choice—and they were very frightened! How in the world had this happened?

Long story made short: the builder had attached all the downspouts from the arena and barn roofs which amounts to about 1500 square feet of roof surface area, and therefore a LOT of rain water run off, into one common, underground drainage pipe that was only 3 inches in diameter. A three inch diameter pipe cannot possibly handle the massive amount of rain that was coming off all the square footage of that roof, and so it backed up. Then, when the water backed up in the pipe, it had no where to go but to back up first under the ground in front of the barn, and then further, to flow backwards into the barn and subsequently under the stall floors. Because the stall floors are made of stone dust, the sudden confluence of water simply mixed with the stone dust and instantly congealed it into a tight mass, shrinking the stall floor substrate, and causing it to collapse. Yipes!

As you can imagine, this was quite a huge problem. Not only did we have frightened horses to deal with who were unable to reach their food and water, and were afraid to move, but we also had to quickly cut off all the downspouts from emptying into the underground drain and then fix the collapsed stall floors fast. Of course the day this happened had to be a Friday which made getting a delivery of stone dust to refill and level the stalls difficult and desperate. Getting a delivery of stone dust was further complicated by the cold weather factor which had frozen the outdoor storage piles of stone dust at the suppliers making it almost impossible for the supplier to even scoop up a load of stone dust to deliver to us. Then, once we finally had stone dust, it took us hours of back breaking work to fix all the stalls, putting one horse at a time into the arena for long periods of time as we fixed each stall. Most amazing to me though was the trust the horses had to display in allowing me to lead them over the stall "gulch" in order to exit their stalls! Horses are indeed endlessly impressive and wonderful!

By the way, the reason Alex wasn't suffering the same frightening situation was simply because the back flow of water hadn't yet reached his stall when I discovered the problem, and, therefore, his stall floor was still flat and level.

<u>LESSON #44:</u> Find your <u>hay farmer</u> now. The three most important people in your life once you are on your farm will be your hay farmer, your farrier, and your equine veterinarian.

Well, I'm sure you'll remember that the soy bean farmer whose crop our first excavator ran over was the #1 choice for the best hay in the area. We actually did try and call him and he actually did refuse to sell us hay. Amazingly enough, the secretary of the first excavator's husband was also a "horse" hay farmer and mentioned to my husband that if we were interested, her husband could be our hay supplier. Well, by the time we needed the hay, we were no longer in good graces with Mr. Excavator #1 and the secretary informed us that her husband would only sell us 100 bales of the 900 we needed and that we were never to come back again. Plus, we were informed that we would also have to come and get the hay ourselves and then stack it ourselves.

I think maybe you're getting a feel that maybe there's a kind of "closed shop" attitude in the farming community? And, you'd be right!

The hay farmer the local mill recommended was a gnarly old farmer dude who would only sell us 200 bales of hay even though he had a hay barn full of hay. He claimed that the rest of the hay in the barn was "sold" as such, but that if we needed more hay later in the winter, that he'd see how much of it might become available.

Do you hear "price game" in those words? And you would be right again!

Apparently that's a fun little game a lot of hay suppliers play wherein they limit your supply in the cutting season so that you are desperate in the winter season and therefore willing to pay whatever outrageous premium they decide to charge you when you need to get more hay.

This farmer also would not deliver the hay. I should note here that there is a huge difference between hay farmers who deliver and those who do not. Pick up one hay bale. It's heavy. Now imagine picking up, and then stacking, 900 hay bales in stacks that are each 14 to 16 feet high. Can you do that? I sure can't! In fact, it's difficult and dangerous just to pull down the bales to feed to the horses!

A good hay farmer will sell you fresh cut hay, and deliver it to your farm freshly bailed from the field, and then stack it in your hay barn. Yes, there is a relatively small charge for the delivery and stacking service, but worth every penny. It is about 25 cents more per bale to have it delivered and stacked. We were paying about $6 per bale delivered and stacked.

A good hay farmer will also freely show you his bales (if you have to buy off season from his storage facility) and readily bore into any bale to show you it's freshness factor including how green it is (many bales look pale in color on the outer layers due to sun bleaching when stored), and that it is mold free. He will also freely discuss your horse's nutritional needs and the differences between his cuttings so as to match his hay to the needs of your horses. He will also freely inform you of the mixture of grasses in the bales and show them to you. He will also be happy to weigh the hay bales for you.

Some hay farmers will only deal with a few large farms that use thousands of bales of hay per year. Other hay farmers prefer dealing with several small farms because it gives them more security of return customers. In other words, if the farmer is only supplying one or two big farms, and the farms go out of business, then the hay farmer will go out of business too. However, if the farmer has 20 small to medium customers and only one or two farms go out of business, the hay farmer's business goes on without a glitch. But hay farmers also have their regular customers and if you aren't one, you aren't going to get the kind of service you want or the number of hay bales you want. They are a decidedly finicky group to deal with.

Also, the farmer will generally expect you to pay before he actually unloads the hay into your barn, unless you are a trusted, repeat customer. The reason for this is that there is one of those little laws that exists that outright screws the hay farmer if he delivers the hay to you and then you decide not to pay. It turns out that in many states, once the hay is offloaded into your barn, it becomes "your" hay, and therefore, technically and legally, you don't have to pay the hay farmer. So, expect your hay farmer to ask you to pay the bill before he actually offloads the hay!

But, now, in the meantime, in our case, we only had 300 of the 900 bales of hay we needed. Clearly that wasn't enough and I was getting desperate.

So, I started asking everyone now, including our workmen, and even Mr. Bobblehead, if they knew of any reputable hay farmers. Well, it turned out that Mr. Bobblehead and our "de-rocker" landscaper both came up with recommendations for the same hay farmer. So, we called the hay farmer, and he not only had great hay, but he delivered and stacked the hay at the above mentioned, reasonable cost. He also took our order for the next season's hay and became our regular supplier of hay.

Knowing the **nutritional analysis of your hay** is also important. Yes, your hay farmer will give you his advice on the feeding of his hay according to his opinion of the richness of his hay. However, to know the nutritional value of the hay for certain, you can find out by having an actual analysis done of your hay. In most states, the state land grant university with an associated agricultural college provides a service in which they can conduct a complete hay analysis for you at a reasonable cost (about $50 per sample). You simply provide a core sample from several of the bales of the same type of hay and send it to them in an envelope or deliver it in person, and they will do the analysis for you and mail you the results.

Core samples are taken using a core sampler tool which can also be purchased thru the university. The sampler is relatively costly but will last forever (about $45). The sampler is able to bore into the core of the bale to allow an accurate analysis.

Private labs are also available to do hay analyses and can be found by contacting your county agricultural agent. County agricultural agents are great sources of information and having your agricultural agent's phone number on your cell phone so you can call whenever you have a question of concern is a good idea.

Once you know how rich or poor your hay is, you can determine how much to feed, what type to feed, and how much supplemental grain or other additives will be needed for your horse's diet and overall good health.

As long as we are talking about hay, you should understand what the **cuttings of hay** mean. I will base this on my hay farmer's fields which are a mix of timothy and alfalfa grasses which is the customary combination of grasses fed to horses and on the typical four season climate where I live.

NOTE: I understand there are states in the country such as Missouri where 6 or 7 cuttings of hay per year are available, and some other states such as Arizona where specialized farming practices are used that yield as many as 14 cuttings of hay per year!

In the late spring (late June), there will be a first cutting of hay. This first cutting of hay is going to be the coarsest of the hay cuttings because it contains stalks and seed heads left over from last year's cutting and has overwintered in the field. It will also be the lowest energy hay, and can therefore be fed the most freely of the cuttings. However, horses with poor teeth—particularly older horses--will not be happy about chewing this hay. Also, horses aren't very excited about the "flavor" of this hay, and will resist eating it if they have a choice. A lot of horse people actually prefer to only feed this lower quality hay and use supplemental grains and additives to meet the varying nutritional needs of their horses depending on the horse's work or show load, and stage of life. That is not necessarily the happiest situation for the horses however.

Horses prefer - just like us - to have a variety of foods. To horses, that means having a variety of hay to eat during the day. Having such a variety would be more similar to their natural state of grazing over huge distances during the course of the day. Therefore, it is desirable to have second and third cuttings of hay available for your horses in addition to the first cutting. The later cuttings will also provide you with a higher energy and more nutritious hay to feed your horses to mix with the lower energy, 1st cutting hay.

NOTE: Never feed all high energy hay.

The second cutting of hay will occur in late July to mid August. This will be the softest, greenest, richest, and easiest to chew of the hay cuttings. This hay will have a predominance of alfalfa in it making it energy rich. Horses LOVE this hay! However, due to its richness, this hay must be carefully fed and managed.

NOTE: Horses get quickly "spoiled" about wanting the "better" hay. I have circumvented this problem by feeding the horses the 1st cutting hay as their night hay.

You will also note that there is a difference in weight between the 1st and 2nd cutting bales of hay. The second cutting hay bales will be denser and heavier than the first cutting bales because they will contain more alfalfa.

The square hay bale should look and feel like it is tightly packed. Don't be shy about weighing bales before you purchase them if you want. Weighing bales is a common pre-purchase practice. In general, a square bale of hay should weigh between 40 to 50 lbs. A square bale of grass hay can weigh as little as 38 to 40 lbs., whereas a square bale of all alfalfa hay will weigh between 45 to 50 lbs. If a square bale weighs significantly heavier—in the range of about 65 lbs.—you should suspect that the hay has too high of a moisture content in it from being baled too soon (not being allowed to dry out in the field properly before being baled). Moisture ridden bales will turn moldy, and we have discussed the danger of moldy hay to horses. If you buy hay that has been stored for a couple of months, you may find that the exterior of the bale is pale in color. Such pale color is normally just a result of surface sun bleaching, and if you core into the bale, you should see fresh looking hay.

NOTE: A horse eats 2.5% of his body weight in **"total"** feed (**grain plus hay**) per day. The **minimum** amount of roughage (hay) per day that a horse **must** eat is 1 lb. per 100 lbs. of body weight. It is preferable that a horse eat 1.5% to 2% of his body weight in roughage per day. Whereas you can't really over feed a horse with hay, you can over feed a horse with grain. A horse should **never** eat more than 1.5% of his body weight in grain per day*, and he should never be fed more than 5 lbs. of grain at one feeding. Also because a horse's stomach is small, his total feed should be split up evenly between 2 to 3 feedings per day.

*A veterinarian may prescribe exceptions to this rule as for example in an attempt to bulk up a senior horse with specialized senior feed. However, any exceptions to this general rule must be made ONLY under veterinarian advice and supervision.

If the weather is warm and it rains just right, there may be a third cutting of hay in the early Fall (September). This cutting is not reliable enough to base your hay supply on however, because there frequently aren't any 3rd cuttings of hay due to the weather. This hay is also very rich hay. How rich each cutting of hay actually is must be analyzed in order for you to know exactly how rich the hay is, and allow you to adjust the feeding of the hay accordingly. The 3rd cutting hay is also soft and tasty which makes the horses very happy to eat this hay as well.

<u>NOTE:</u> Although it is desirable to have first, second, and third cuttings of hay in your barn, all three cuttings should come from the same field. This will lessen the chances of gastric upset in your horses. I'm sure you already know that horse digestive systems are VERY sensitive and are easily prone to colic.

So the "perfect" day for a horse would be a mix of 1st, 2nd and 3rd cuttings. Perfect days don't usually happen, however, and so some kind of a mix of 1st and 2nd or 1st and 3rd should work out happily for all. I use a two to one mix of 1st to 2nd or 1st to 3rd for my horses.

The bottom line is that whereas all the hay comes from the same hay field, the different cuttings of hay will all look and be different depending on the time of year they are cut and baled.

<u>NOTE:</u> You need to have a scale to weigh your hay. Weigh a flake of each type of hay so that you know exactly how much weight in hay you are feeding your horses. Generally speaking, a flake of hay will weigh between 2.5 and 3 lbs. Remember that different types of hay have different energy levels, such as alfalfa hay, and 2nd cutting hay having high energy levels, while grass hay (hay that does not contain alfalfa) and 1st cutting hay having low energy levels. Remember too that horses must be fed a minimum amount of forage per day—1 lb. per 100 pounds of body weight. As such, **<u>feeding hay to horses requires a combination</u>** of the correct minimum weight of hay (forage) that needs to be fed per horse per day according to his body weight, PLUS the correct ratio of energy levels contained within the various types of hay you are feeding that correspond to the horse's stage of life and activity level. Begin with a mix of hays you think will provide the appropriate energy level for each horse, while being sure to feed at least the minimum weight of forage, and then observe your horse's energy level and adjust the **<u>type</u>** of hay fed accordingly. And, lastly, it is always better to err on the side of feeding a little too much hay to your horses than to feed too little hay per day.

<u>NOTE:</u> Be sure to bring some hay from your old barn to your new barn in order to transition your horses from the old hay to the new hay. You should transition the horses gradually over a week by slowly adding the new hay to the old hay. Start with about 1/8th flakes of hay—new hay to old hay —and then gradually add more new hay each day over the first week of feedings.

NOTE: I did not opt for putting hay feeders on the walls in the horse stalls, nor do I use hay feed bags mounted on the stall walls. Some people make the argument that hanging hay is cleaner for the horses in that the hay is lifted up above their soiled shavings, thereby keeping their hay clean. The opposing argument is that horses naturally feed with their heads down, and, therefore, having the hay on the floor of the stall is the correct way for a horse to eat his hay so that his entire gastro/intestinal system works properly. As far as I have seen, the horse with a wall mounted hay feeder basically pulls out the hay from the feeder and lets it drop down to the floor, and then eats the hay off the floor, thereby eating the hay off soiled shavings anyway.

But now it's finally time to return to our seriously sidetracked barn planning. I will share with you my barn design and my thinking about why I did certain things.

It turned out that our first barn plans were grandiose. We didn't think the plans were grandiose when we drew them up. We only came to realize they were when we started getting back builder quotes that were breathtakingly grandiose. So, we started cutting back our barn design, and eventually came to cut our barn design way back.

LESSON #45: The reality of operating a commercial facility may not be as attractive as it seems in the planning stage.

After so many years in riding, we thought it would be an easy step for us to build and run a **commercial facility**. After all, the sport of riding can be very social and I liked that part of riding very much. Having my riding friends over all my years of riding had provided me with many happy times and memories.

So our first plan was to build a barn with enough stalls to have boarders and have a viable income from the boarders that would allow for a nice yearly income. We planned a center entrance barn with a grand foyer and horse stall wings running to the left and to the right off the center entrance. We calculated that having a minimum of 16 stalls would allow for a nice income on the farm without having too many boarders. My husband was confident that I could easily maintain those 16 stalls by myself which would give us a real start on money flow before we had to hire in any help.

Wrong.

LESSON #46: If you plan to have a commercial facility, you must plan on having hired help.

I can tell you now, but I couldn't have told you back at the planning stage, that mucking stalls is both labor intensive and time consuming. It takes me at least a half hour per stall to muck it out. It also takes all my strength and energy away by the time I have mucked out just three stalls. In addition, after a few months of lifting the heavy muck rake, you will find that you will begin to start suffering physically with various repetitive strain injuries in your hands, hips, and knees.

Plus stall mucking time, when you are running the farm on your own, is only one part of your time. You will also be the one carrying the shavings bags over to the stalls, opening them and distributing them. You will also be the one cleaning out the horses' feed and water bowls. You will also be the one getting the hay from the hay barn, and then distributing it. You will also be the one getting the grain out of the grain bins and graining the horses. You will also be the one doing the turnouts. You will also be the one grooming the horses. You will also be the one sweeping the aisles. You will also be the one wetting down the arena and dragging it. You will also be the one ... well, I think by now you get the idea.

So, considering how much work you will have to do on your own farm, ask yourself if you will have any time or energy left to actually ride your horses. The answer quite frankly is not much, if any at all. You may even find that after all the time and labor you have to do in just caring for the horses that you are just sick of being in the barn and you won't even want to ride the horses.

But on the other hand, there are several issues with hiring people. In the first place, ask yourself who you would hire. Then ask yourself what the quality of their work would be. Could you fully trust your workers? How would you fire them so that they weren't bitter and retaliatory? You need to ask yourself all these types of questions before hiring in help, and you have to feel comfortable with your answers. In the end, you must also ask yourself if by hiring in help that maybe all you are doing is trading off regaining your physical strength only to have your mental health destroyed by unending problems involving your workers.

You remember the woman's driveway that runs down the middle of our property? Well, her husband is unfortunately a stroke victim and confined to a wheelchair and overall disabled. As such, she has hired help come in to assist her in the care of her husband. Not being the most easygoing person in the world, the woman has a constant turnover of caretakers—about one new caretaker every month.

One day in the spring, her current employee stopped to chat with me while I was working on the garden by the sign at the entrance to our farm. She very considerately had thought maybe my car had broken down and that I needed help. She had a thick Spanish accent. She was very pleasant and laid back and friendly in her Spanglish. Eventually in the conversation she asked if perhaps next week she could drive down our driveway and see the horses and barn because she had been raised on a farm in Costa Rica. I said that would be fine and that I would expect to see her in the morning of either that next Tuesday or Thursday which were the days on which she wanted to visit.

When she came to visit the next week, she was still very pleasant. However, she was now also surprisingly aggressive. As we walked through the barn, she would animatedly begin dusting anything we stopped next to. Meanwhile, I should mention that there is never a shortage of things to dust off in a barn. In fact, there is a very short return dust cycle to objects in the barn that occurs each and every time a bag of shavings is opened, a stall is mucked, or an aisle is swept. Dust clouds unendingly erupt in barns. So if being hired to dust the barn was this woman's goal, it was a no win goal because I certainly wouldn't hire anyone to chase dust around my barn.

Next, the woman began giving me a long list of reasons why she was employable by me on my farm even though I had noted that the blood had clearly drained from her face when she actually saw the size of the horses in their stalls up close and personal as versus driving past them in the pastures in her car. But she was quick to mention that although she had been raised on a pig farm, that her mother had had a horse "one time." Then she proceeded to kind of summate her personal history by noting how she and ALL her brothers were trustworthy and hard working. And then finally she concluded her sales speech by mentioning that she was now available for hire in that our neighbor had fired her from caring for her husband.

Oh, boy. Here I was in my barn with a very aggressive woman pressuring me to hire her (and ALL her brothers, I guess) when I had no interest in hiring anyone at that moment. But my next immediate thought was that even if I wanted to hire someone like her, how in the world would I begin to make sure in the first place that she wasn't an illegal immigrant? And further, that if this woman was this aggressive during her sales speech, then how in the world would I control her as an employee or potentially get rid of her if I needed to fire her?

Right then and there, I decided that I would rather deal with my various aching body parts for a little longer than have this type of worker around the farm.

<u>LESSON #47:</u> Commercial facilities DO have to meet code and are therefore more costly to build and are subjected to the close scrutiny of the building inspector during construction, and then eventually to the close scrutiny of the tax collector!

One of the commercial barns we visited before starting our construction had the owner still "steaming mad" about a very expensive fire door the building inspector had forced him to install between his attached arena and his barn aisles. It was a HUGE kluge of a door like an industrial strength, metal garage door with a medieval cumbersomeness and an ENORMOUS motor drive attached to it. Not only was the motorized addition a huge spook causer, it was clumsy and ugly, and so costly that the owner would only hiss at the thought of discussing the price of it.

As another example of what would be required if you built a commercial facility from the start is the need for a totally ADA wheelchair friendly environment that includes everything from easy access entrances into the barn and stalls to easy access bathrooms. Such things add a lot more cost to your building project.

Of course there are ways around the situation. There are grandfather clauses in some areas that allow you to turn an existing private facility into a commercial facility after a specified number of years of being a non-commercial facility without having to meet any current commercial codes. Then there are loopholes in laws in some places that allow you to have as many family members of your immediate and extended family to board and ride at your facility without the facility having to be deemed a commercial facility. That is what the owner of one new barn we visited was claiming in order to avoid meeting code at her facility. She made a point of introducing everyone to us in her obviously commercial facility as being her relatives after which she followed with a quick wink and a smile.

Unfortunately, I don't know how to find out about the existence of these loopholes and grandfather clauses for sure. You could try your local zoning office, but the level of secrecy and lying about informing you of these kinds of situations is breathtaking. I didn't try to find out any "angles" when I built to allow me to build a non-commercial facility and then run it as a commercial facility. However, I would recommend that you try to find grandfather clause information in case you ever want to become commercial. I would suggest you do that by first traveling around to various barns and asking what they know about such clauses and loopholes and then following up that information with research. Once again of course, you can hire an attorney to research it for you if money is no object.

<u>LESSON #48:</u> Even if you plan a commercial facility and think it will bring in profitable income, if you have never before run a commercial facility, you may be disappointed. To maximize the probability of success, perhaps you should consider a two step process in which you first operate a private facility and then convert it into a commercial facility.

I had heard for years from several sources that a horse farm was a sink hole for money and nothing else. One farrier had told me that one of his customers, who had built a beautiful facility and had a full barn of customers, told him that building the farm and running it was the biggest mistake of her life. I just quietly stood and listened to the farrier as my mind scoffed off the barn owner's quoted remarks as somehow bitter and unrelated to me and my future farm. Then even our financial investment counselor started talking about someone who had built a horse farm and tried to run it as a profitable business. He summated his story by telling us that everyone who was "in the know" knew that the quickest way to go broke was to try and run a profitable horse farm.

But I still thought I knew better. After all I was paying a small fortune per horse at the abysmal horse facility where I was boarding. So why wouldn't people want to pay a premium price to board at our beautiful new facility?

Well, I had it half right. There was no shortage of people who wanted to board at our barn once it went up and even while it was in the process of going up. The problem, however, was that most of these prospective boarders not only didn't feel they should pay a premium to board at our barn, they wanted to board at our barn essentially for free! They wanted to treat our farm as a co-op!

People are amazing in how there is no end to their sense of entitlement! Be warned. The good news is that there are still realistic, and real paying customers in the equestrian world.

Meanwhile, once we started getting horrifically huge quotes on the commercial barn we had originally planned, we started considering other types of barns and smaller ones to be more in line with our budget. We also decided at that point in time that we would change our original plan and approach having a commercial farm as a two step process. The first step would be to build and operate a private farm that could be easily and inexpensively expanded into a commercial facility. The second step would be to expand the farm and begin to take in boarders who would be real paying customers. Having a core number of real customers would then permit us to hire trustworthy help. Then, once we had a core customer base and trustworthy help and suppliers, we planned to further expand the revenue of the farm by operating it as a show facility.

Nevertheless, back to considering the outrageous quotes we were getting, what was so amazing to us was that no matter how many times we changed the size or style of the barn, the quotes that came back were relatively close to the original horrendous quote.

As an example, here are some of the various style barns we considered and the **quotes** we received:

(a) A center entrance barn with stall wings to the right and left of a large center entrance - 16 stalls - metal exterior (pole barn) - tongue and groove pine interior -- $360,000.00;

(b) A-frame pole barn with an arena in the center and 16 stalls lining the edge of the arena, separated by an aisle (8 stalls on each side of the aisle) -- $240,000.00;

(c) Four stall, <u>wood</u> barn with tack room and grooming stall (no arena) -- $250,000.00;

(d) Four stall, pole barn with tack room and grooming stall and second story, attic hay storage -- $265,000.00;

(e) Eight stall, center entrance pole barn – no interior finish -- $320,000.00.

This was incredible pricing. Remember these quotes were "first passes" at numbers. By that I mean there were many essential items not included in these quotes. It did not include full lining of the aisle walls with pine tongue and groove finishing (beyond the tongue and groove of the actual stall walls) or quality exterior windows and doors, or quality stall fronts, or appropriate levels of lighting, or a cemented aisle floor, and on and on. No, these were only "entry level" prices and they were outrageous. Remember too that in four of these quotes, there wasn't even a place to store hay within the barn. We would still have to build a separate hay barn at an additional cost. Also, it was only one of these quotes that had an indoor riding arena. You may be thinking now that it's okay not to have an indoor arena, but once you are living on your farm and the first winter or wet season, and the first hunting season present themselves to you, you will be very glad to have an indoor arena of any size!

Once we got these initial quotes we decided that we would rather have a smaller barn in order to have an arena. We also determined that we would first find a reasonably priced, quality pole barn company rather than trying to build an overpriced wood barn, and then second, find reasonably priced barn products for customization of our barn (like European stall fronts) by shopping online.

In time, we did indeed manage to find a great pole barn company whose quote allowed us to build both a barn and an arena and to finish them off to a reasonable degree with the kinds of basic things that were important to us for our horses as well as for our daily horse management.

So now let's talk about your barn plan. You already have a rough idea of how many stalls you want or need plus the service areas for your barn which you used for your original estimated barn size on your approved site plan. Of course I don't know what you have in mind or have drawn on your paper and I obviously cannot directly help you here with your exact plan. What I can do, however, is talk about all the things I did and why I did them as well as the information I "discovered" along the way. Again, I offer all my thoughts and actions to you only as food for thought in your personal planning. You certainly don't have to agree with me. You may in fact find that you want to do things completely opposite of what I have done. Nevertheless, I do have an attractive, functional barn built within a reasonable budget that can be used as a template against which you can upscale or downscale your project according to your personal budget and/or needs. I sincerely hope that my experience will be helpful to you in your overall barn planning.

Oh, there's one last thing to consider—or you could think of this as being the first thing in the beginning of your barn planning to consider. I'm talking about horse manure or **muck**!

The reason I bring this up now is that muck—what to do with it, where to store it, and where to get rid of it—is an overwhelming point of concern for horse farmers.

The traditional way to dispose of muck is to use a muck spreader. Muck spreaders come in a variety of sizes from conveniently small enough to be easily pulled behind a Gator, to commercial duty, large sized ones to be pulled by a tractor. Muck is pitched either directly from the stall into the spreader or dumped into it from a wheelbarrow pushed up a ramp. Then the spreader is pulled by the tractor or Gator out into fields where the muck is spread across the ground.

This system is all well and good and traditional and ecological, but it does have a few inherent problems.

The first problem is that there may be ordinances about spreading the muck that disallow you to spread the muck on your fields. If your farm is in any sort of a subdivision, there may be covenants preventing you from spreading muck on fields.

The second problem is that you just may not have enough land on which to spread the muck. Remember that horses produce a LOT of muck—every day! (We fill a 12′ x 12′ x 4′ high muck rack each week with muck from just 3 horses!) But even if you have a large area of land on which to spread the muck, over time it may become overfilled and, at least for a period of time, become unusable.

NOTE: Do not spread muck on a pasture that is actively being grazed by your horses—it's just flat out unhealthy for the horses.

The last problem is that by spreading muck on open fields, you are providing a perfect breeding ground for **flies**—thousands and thousands of flies. However, there are various products available to control flies.

One preventative product that you can use is a feed supplement that passes through your horse's body and breaks the fly life cycle by preventing the formation of fly larvae's exoskeletons when they molt. The flies die and cannot propagate. Personally, I don't think that feeding your horse a daily fly poison is a great idea, especially over time.

Another solution to flies is available and uses a predatory insect (a miniature wasp) to attack fly pupa killing the immature flies before the flies can reproduce. By spreading the predator insects near manure, the fly population is limited and minimized either completely or at least to a tolerable level. This product requires monthly application during warm months—and therefore, potentially year round in warm climates—and it is NOT cheap. Although I have used this product and I have had some success with it, I have to say that the product didn't really get a fair test because it was up against an overwhelming challenge which I will share with you.

It happened almost immediately after we had set up the fans our first summer. We were feeling happy for our horses that they were not only enjoying the breeze from the fans, but were enjoying the fans' ability to disperse the few flies that were in their stalls. But right after that brief moment of satisfaction, we also noted an offensively pungent odor undeniably wafting toward the barn which the fans were clearly serving to amplify. Wondering what in the world the source of the stench could be from, we looked out of the barn windows. What we found to our horror was a farmer dumping basically raw manure from a huge piece of farming equipment his enormous tractor was pulling around the adjoining acreage to our farm. Apparently the lovely woman who had sold us our land had leased this adjoining 20 acre chunk of land to this farmer.

What we subsequently learned was that a cow farmer had leased the land to dump all his cow waste onto it. The bridge near his farm was being repaired and had blocked direct access to his traditional dumping ground. As such, he had leased the property adjoining ours to dump all his liquefied cow waste until the bridge repair was finished—which would be a matter of months!

I can't begin to explain to you just how overwhelming the fly problem was that this waste dumping created on our farm. It was totally disgusting. Worse, the situation for the horses was just pitiful.

So, how bad was the fly problem? Let's just say that by the time the farrier drove through the entrance gate into our farm in his white truck and arrived at the barn door—approximately 130 feet—his entire truck was COVERED with flies. As he exited his truck, the farrier slowly drawled out in his classically laid back cowboy style, "You know, I'm thinking here, that you don't just have flies—you have a fly problem!"

This was the level of challenge the predatory fly product was up against. Although the flies were decidedly reduced on our farm with the use of the product, we still had a problem as such. I plan to give the product another try this year as well as to begin using the product in the beginning of the spring season so that it can have a better chance of grabbing hold of the fly population before it becomes overwhelming. But my bigger hope is that the bridge gets repaired and the cow farmer, and his flies, all disappear!

UPDATE: The good news is that the predatory fly product is a winner! This year I began the predator program in the early spring as recommended by the manufacturer, and spread the product in several perimeter areas of the farm as well as in the muck rack area and in the barn. As mid June approached and we still didn't have a fly problem or anything more than an occasional fly at the farm, I began to wonder if the lack of flies was due to the predator product or if it was just due to this being a low fly population year. Also of note in my thinking was that the adjacent farm land had not been leased to the cow farmer this year such that I wasn't sure that his absence was the reason for the low fly population. So I didn't know for sure what was up with the lack of flies on our farm until yesterday when I got my monthly shavings delivery. The deliverymen couldn't stop marveling at how there were NO flies at our farm. According to them, this year had turned out to be one with an overwhelming fly infestation at every single barn where they had delivered—except for our farm. They asked us why it was that we didn't have flies. We told them that we were using the predatory fly product and assumed it was because of that. They in turn told us that they couldn't wait to return to their store to encourage the manager to order in the fly predatory product for use on all the farms in the area—especially the ones where they made their deliveries!

Another "green" way to deal with muck is to compost the muck. However, composting your muck will require a LOT of land. You will be creating a huge pile of waste on your property in one big mound. This mound of waste will fall subject to all the aforementioned issues of spreading the muck on your property including local ordinance restrictions that may forbid compost piles, all the way to the issue of breeding overwhelming fly populations. However, if allowed, you could opt to compost at least "some" of your muck, and use it for enriching gardens and lawns at the completion of the composting cycle.

Another way to remove muck from your farm is to have a commercial trash service dumpster put on your farm and then have that trash service remove the muck from your property once a week. We were charged $115 per month for an 8 cubic yard dumpster about 6 ½ feet wide, 7 ½ feet deep, and 6 feet tall that was emptied weekly. In some areas of our country, your trash removal people will have no problem renting you a dumpster and allowing you to mix your regular household trash with barn shavings and manure.

However, in other areas of the country they consider mixing the muck and garbage as ecologically incorrect. There is, however, an answer to this problem in that even in these areas there are some trash removal services that will rent you a dumpster that is to be solely used for the muck. My understanding is that this type of "ecologically correct" dumpster is lined with the equivalent of a huge plastic garbage bag. I really wouldn't be able to tell you why lining a dumpster with plastic makes muck suddenly ecologically sound, because I am chastised for putting my kitchen garbage into plastic bags for disposal. I would also be hard pressed to explain why the bacteria decomposing my kitchen garbage is any "healthier" than the bacteria breaking down my horses' waste. After all, if push came to shove and I had to do a comparison between my kitchen garbage and horse waste, I would have to first point out that the dogs regularly eat the horse poop and remain healthy, happy, and worm free. But should the dogs eat a piece of uncooked chicken skin from my kitchen waste that I threw out about a week ago, I don't think they'd be feeling so good—or perhaps even survive. Nevertheless, that is the way things are.

There are some places where you can get specialized muck removal services that involve heating up the muck to make it environmentally friendly. This usually involves only a once a month pick up of the muck which creates a larger temporary muck storage problem. The cost of such specialized removal is about twice the cost of a dumpster rental—affordable, but not "cheap."

LESSON #49: If you are going to have a dumpster, and you don't plan ahead as to where to place your dumpster, then you may have to set it right in the middle of your beautiful farm setting and ruin the beauty of the overall look of the farm.

The point of discussing "removal" of the muck from your property now is that if you can choose, and you decide to choose to have a collection dumpster and removal service, then you will need to plan where you want the dumpster to be placed. You will also need to plan for driveways and turnarounds that will accommodate the large size and heavy weight of a dumpster garbage truck. When determining where you should place your dumpster, I would suggest you select a spot rather "hidden" from view because there is nothing "pretty" about a dumpster. Unfortunately for us, this was one item we missed in our planning, and, as such, our dumpster is in full view, and rather centrally placed, and looks flat out ugly!

You should also consider right from the start where your temporary muck storage will be if you will be using a commercial waste removal service, and depending on how often it will be picked up, how large a temporary storage area you will need. You will need a **muck rack** where you will dump the daily muck and accumulate it before it is put into the dumpster. You will need your muck rack placed where it will be easy for you to access from the barn—regardless of the weather and driveway conditions—yet "out of view" and a good distance away from the house or barn. If you have a dumpster, you will be using the front end loader bucket of your tractor to scoop up the muck and then drop it into the dumpster once a week before pick up (if hopefully you have once a week pick up). As such, having the muck rack and the dumpster close to each other will make the muck transfer process (from storage rack to dumpster) much quicker and simpler.

Painting the cement block walls of your muck rack with a protective, waterproofing paint is a good idea. Moisture contained in horse waste can cause the mortar between the cement blocks to crack over time if you live in a climate where repeated freezes and thaws occur.

Photo #29: Muck Rack

Now that we've covered the muck subject, let's finally start planning your barn with a **basic barn plan**.

Chapter Six

Horse Barn Planning - Second Phase

NOTE: Please remember that **absolutely everything in your barns, your arena, your aisles, the horses' stalls, the pastures, and every area in between those other areas, MUST have smooth edges and surfaces, without any protruding points or objects anywhere** or your horses will injure themselves. Horses love to rub and will do so at their own peril, totally ripping themselves apart on any protruding objects or rough edges or surfaces. They will also run into protruding objects and seriously injure themselves if you don't take the precaution of removing them for your horses.

Clearly the first thing to consider in your barn plan is the **number of stalls** you will need. The concept of need here is planning for how many horses you currently have and need to house, plus the number of potential future horses you would like to have. Need can also mean how many stalls you need for a commercial facility to achieve a particular financial goal and/or to house your current and future horses. Either in a private or a commercial facility, I would recommend planning for at least **one extra "empty" stall**. You will find many reasons present themselves to use that extra stall and you will be very glad to have it. Just thinking simply, as a point of example, ask yourself where you would put a horse if something went wrong with his stall (like the stall floor collapsing!) and you couldn't get it repaired for more than a day or two?

Then, after you determine the number of stalls you need to have, you must decide what **size the stalls** should be.

Basically stall size depends upon horse size. If you have ponies or small horses that are close to being a large pony size, then a 9′ x 12′ stall is sufficient for each pony or small horse. However, if you have full size horses, then the minimum stall size is 12′ x 12′. If you have large dressage or Grand Prix horses that are more than 16.2 hands and weigh more than 1200 pounds, then you should seriously consider having 12′ x 14′ stalls. Larger stalls are also needed for stallions, and as foaling stalls.

Chances are that if you go in and measure the stall size at commercial facilities, either where you are boarding or to one you have access, that you are going to find that none of the stalls are larger than 9′ x 12′. In fact, you will most likely find that the stalls at commercial facilities measure even smaller than 9′ x 12′.

Why do commercial horse facilities generally have stalls 9′ x 12′ or smaller? First of all, rather obviously, with smaller stalls the facility owner can fit more stalls into a prescribed barn size which in turn translates into more horses that the owner can board. The more horses the owner can board, the more revenue there is to be gained. But the additional, "hidden" reason why smaller stalls help make more profit for the facility owner is the cost of shavings. Shavings are expensive. Having smaller stalls means having fewer shavings to buy and replace per stall.

If making money off horses is your primary focus, then you will be drawn to building inadequate, smaller stalls. However, if you take a minute to look at the full sized horse in a small stall, you will see why it is not good for the horse.

The full size horse will only have just enough room to circle in a 9′ x 12′ stall. Big horses won't really have enough room to easily circle, if at all. Within that circle, the horse has to eat his hay, as well as urinate and defecate. Also within that circle he will have to lay down when he is tired. Guess what he's going to be laying down in when he's tired? He also has no "other space" within his stall. He can only stand where he circles and relieves himself.

<u>Surprise:</u> Horses prefer not to sleep where they eliminate—just like us—imagine that!

When the stall size is correct, a horse will select an area for urination and defecation and predominantly eliminate there. He will have other areas in which he will "stand sleep" and in which he will lay down. He will make designated places where he puts his hay to eat. He will not be inclined to agitatedly circle in his stall because he does not feel cramped. He will be a much more psychologically sound and happy animal in the correctly sized stall. He will also not be standing full time in his urine and feces and therefore he will not be as inclined to get thrush.

If your horses are going to be turned out all day, every day, then you can "get away" with smaller stalls. However, smaller stalls are never right or desirable for the happiness and health of the horse.

I briefly housed one of my horses at a private facility while he mended a pulled suspensory ligament. The owner, who was very gracious to allow me to briefly board at her facility, continually apologized to me for the smallness of the stall my horse was housed in, particularly compared to the stallion/breeding stall her mare was housed in. I should note that this was one of those situations where even though the stall was inadequate in size (10′ x 10′), the horses were turned out all day and so the situation, in the short term, was tolerable. Anyway, what caught my attention during my horse's stay at her barn, and made me remember the owner's remarks, was regarding her mare in that big 10′ x 18′ stall. Her mare was super nasty and the owner had pretty much given up on riding her. The owner was merely using the mare as a brood mare which she actually was when the owner had originally purchased her. The one redeeming quality that the owner kept mentioning to me about the horse was how "neat" the mare was about her stall. She laughed about how she had noted that the mare had divided up her stall into little areas just as though she was a human being living in a little apartment. She noted how the mare had her bathroom area, her kitchen area where she ate her hay, her sitting room area, and her bedroom. The "neatness" of her mare endlessly impressed the owner, and as I mentioned, the owner repeatedly offered this information to me as a reason to admire and tolerate her otherwise flat-out nasty mare.

Meanwhile my gigantic horse was standing in his poop in the neighboring stall and was clearly miserable. I knew my horse was neat and cared about being clean. He was probably the least stained gray horse that ever existed. So as I stood there listening to this owner preach about the wonderfulness of her nasty, old bay mare, I couldn't help but wonder why my horse wasn't as fastidious in his stall as was her mare. My eyes scanned across my horse's urine and feces stained blanket. As I continued to stand there listening to the woman's bragging, I finally decided in my mind that maybe the cleanliness thing was a mare versus gelding thing. Yet, somehow that didn't seem quite right to me based on my horse's usual cleanliness (before coming to her barn), and was why that scenario kept tugging at my mind long after I left her farm.

It wasn't until I finally had my own barn with its 12′ x 12′ stalls that the obvious answer became clear. It wasn't a "mare versus gelding" thing, nor was it a "your horse is better than my horse" thing—it was purely and simply a horse thing! My horse just didn't have enough room in that 10′ x 10′ stall to be able to be fastidious. Now in a 12′ x 12′ stall, he finally can be, and is, completely fastidious.

I know that on paper it doesn't sound like there's that much difference between 9′ x 12′, 10′ x 10′, and 12′ x 12′ stalls. But, trust me, there is! Ideally, all horses should have 12′ x 14′ stalls, in my opinion, but that does get costly in shavings and hard to muck out. I will leave the 12′ x 14′ stalls for those of you lucky enough to be able to upscale.

Now let's consider what you need beyond your basic number of stalls "plus one".

Note: If you have a large commercial facility, I would recommend having at least one empty stall per 10 to 15 horses because the likelihood of needing several empty stalls available increases proportionately with the number of horses housed in the facility. Also the need for multiple grooming and wash stalls increases with a larger number of housed horses at a commercial facility. I would suggest having one grooming and one wash stall for each 20 horses. The wash stalls can be mostly outdoor wash stalls.

First of all you will need a **grooming stall**. Although lots of you have probably groomed in the crossties in the barn aisle, and that's okay if you have to downscale, it is still better to have a dedicated grooming stall. First of all it allows your horse to be isolated from the other horses while being groomed which makes him more relaxed. A dedicated grooming stall also gives the farrier a place to work where he can focus on dealing with the horse "of the moment." It also gives the vet a place to work on your horse if for some reason he needs to work on the horse outside of the horse's stall. Also, if you make the grooming stall in the same dimensions as your regular stalls, and finish the walls with the same tongue and groove pine as in the regular stalls, then, if need be, a grooming stall can easily be converted into an extra stall.

Next you should have a **wash stall**. Your wash stall should also be nice and roomy. I suggest it be the same size as your 12′ x 12′ stalls. Although many barns only have an outdoor wash rack with COLD water, I find it invaluable to have an indoor wash stall with hot and cold water both for washing horses and for washing equipment—especially in the transitional seasons when I (and the horses) would rather be working with water indoors.

With multiple grooming stalls and wash racks, resist the urge to do overly narrow stalls and racks or a shared containment wall line up or quad patterns. Always think about the probability of horses fighting with each other when planning any type of containment structure for working on horses. First, racks or stalls that are too narrow will give the human no place to shift if the horse suddenly and unexpectedly shifts, potentially squashing the human between the containment rail and the horse. Second, it is essential to plan in a human escape path between the grooming stalls or wash racks if they are open style and in a line up or in a quad pattern. The human should always be able to easily duck under the horse containing rails (which should be high enough for an effortless "duck"), and quickly get out of the horse's way; i.e. not have only a path of escape into an adjoining narrow horse stall or rack!

You should have a **tack room** with a locking man door, a **bathroom** that is entered from the aisle, and an **office/lounge/kitchen with a viewing area** looking into the arena. You will need to have a **utility room** where all the utilities come into the barn to then distribute the energy and water throughout the barn, plus contain plumbing fixtures such as water heaters. And don't forget that you will need an **entrance foyer** into your barn as well as an entrance from your barn into your arena!

Above your enclosed rooms—the tack room, office, and bathroom—you should have the builder reinforce the ceilings so that you can have **storage areas up above** those rooms for out of season articles and myriad other items you will find to store. You should also have a storage area in the barn **for your wheelbarrows, muck rakes, brooms, dust pans, and shovels**. This storage area can be as simple as a "nook" that opens to the aisle. You will need an additional storage area within the barn **for bagged shavings** that is easy for the delivery man to access, as well as a storage area **for several days supply of hay** brought over from the hay barn. Then, too, you will need a **storage bin area for grain** and feed supplements.

One last **optional** storage area to consider including in your basic barn plan would be an area in which to **store winter blankets** if you don't want to have plastic storage boxes or show trunks lining your barn aisles.

Now how do you go about envisioning and planning these areas?

Basically you should think of your structure in 12 x 12 foot sections. Conveniently, pole buildings are built on 12 foot framing sections which will provide each of your 12′ x 12′ areas with the strongest support.

Start your plan by lining up your 12′ x 12′ horse stalls, 12′ x 12′ grooming and wash stalls, the 12′ x 12′ tack room, the 12′ x 12′ storage nooks, and the 12′ x 12′ foyer. Just focus first on where you want these areas in relationship to each other. Don't worry about sizes yet. Once you have the basic barn for horses, then add in your other areas, thinking again in 12′ x 12′ sections including the bath, office, lounge, and connector aisle areas.

When you have all the 12′ x 12′ areas placed where you think you want them, now you can begin to change the individual area sizes. Start determining area sizes by dividing up the 12′ x 12′ areas in halves or quarters and then adding or subtracting the leftovers to the adjoining 12′ x 12′ spaces. For example, whereas a 12′ x 12′ office is small, a 12′ x 12′ bathroom is excessive. By placing the bathroom next to the office, you can take 6′ x 12′ off the bathroom and add it onto the 12′ x 12′ office area making the office a roomy 12′ x 18′ area. I should point out that you should keep your aisle width throughout your barn at 12 feet in width.

Your **arena** is planned differently in that the space planning of the arena has additional considerations. The arena requires a large open expanse that has to be supported by either **trusses or steel girders**. Depending upon which type of truss or girder support system you use in your arena, your long wall spacing planning will vary according to the required spacing of the trusses or girders used. As such, the 12 foot spacing used in planning your barn in sections cannot be used in planning for the long walls of your arena.

NOTE: Large commercial projects generally use steel girder framing for their projects. Steel girder framing is very expensive and not ordinarily used for non-commercial horse farms.

The reason the large, open expanse of the arena is an issue has to do with expense. Trusses are prefabricated up to a maximum width of 72 feet with a 3.5:12 roof pitch. That prefabrication width is limited by the bridge height limits of the highway system in our country. Trucks cannot transport trusses that are wider than 72 feet with a 3.5:12 roof pitch, and therefore the limit for the size of prefabricated trusses. This 3.5:12 pitch is the lowest roof pitch that, in my opinion, is still attractive. You can get wider, prefabricated trusses shipped for your arena, but with a much flatter roof pitch. A flatter roof pitch is not only less attractive, but it is less able to allow rain water and snow to slip off the roof. You can also have wider wood roof trusses built on site if you need a wider expanse than 72 feet and want a higher roof pitch, but the cost will be significantly greater than prefabricated trusses, and potentially prohibitively so.

Because I used the prefabricated, wood trusses, the long walls of my arena are based on eight foot sections due to the eight foot spacing between each prefabricated, wood, roof truss. If you plan to use prefabricated, wood trusses, use eight foot sections for planning the long walls on your arena.

Photo #30: Truck Delivery of Prefabricated Arena Trusses

For reference, a full size dressage arena is 20 x 60 meters or 66 feet by 198 feet. A small dressage arena is 20 x 40 meters or 66 feet by 132 feet. A standard size competition arena is 100 feet by 200 feet which is large enough to accommodate a full jumping course.

Our arena is 72 feet by 136 feet. We can fit a small dressage arena inside our arena. We can also put up three jumping lines in our arena. If we use the diagonal jumping line twice, we can have a "full" four line hunter course. Although our arena size is not appropriate for "A" circuit hunter/jumper (h/j) indoor shows, it is fine for h/j practice, h/j schooling shows, and lower level circuit h/j shows, as well as for small arena dressage shows, and western pleasure shows. I should also point out that a lot of people, especially western riders, build arenas that are 60′ x 120′. There is a tremendous difference in size between an arena of 60′ x 120′ and an arena of 72′ x 136′.

You should also think of how your barn areas will relate or **flow** to one another as far as happiness of your horses and your ability to easily work with them. For example, do you want your horse stalls to all be in a line on only one side of an aisle or do you want your horses facing each other across the aisle from each other? (Remember that the horses on one side of the aisle have a totally different life perspective than the horses on the other side of the aisle due to factors such as receiving differing amounts of sun and breeze at different times of the day.) Do you want the entrance to your barn to be centered with horse stalls to the left and right of the entrance or do you want your entrance to be on one end of the barn and then the horses either all to the left or to the right of the entrance? Do you want all your horse stalls on one side of the barn and then all the storage areas on the other side of the barn? Do you want any horses stalled right next to the grooming stall or wash stall or do you want to have separations between the horse stalls and the "service" stalls and storage areas? Do you want separate nooks for each of the various storage units or do you want one large open area for all the storage?

These are the kinds of questions to ask yourself as you sketch out your barn and arena plans. Then to answer these questions you have to really think through exactly how you want to work with your horses. While looking at your barn and arena sketch, actually envision yourself doing various tasks in full detail. Think of yourself taking a horse from a certain stall, walking him to the grooming stall and crosstieing him, getting brushes and tack from the tack room, taking off the horse's blanket and hanging it up, grooming him, then tacking him up and walking him past what kind of rooms, down what kind of corridor, through what kind of arena gate, and into the arena. That's the kind of thinking you should do as you stare at those little 12′ x 12′ squares you have drawn on your paper barn plan.

There are other items regarding flow to consider.

First of all, the location of areas that will require heat, electricity, water, and other utilities should line up with each other to enable the utility services to easily run between these rooms and areas. This would mean that the bathroom, the wash stall, the tack room, and the viewing room/lounge/kitchen areas should all be located near to or adjoining each other.

NOTE: The water lines that run to your wash stall must be run through heated walls so that the water lines do not freeze in winter.

Enclosed rooms and areas requiring utilities can be located across the aisle from each other if the utility can run its service overhead, across the aisle. Running utilities overhead, however, can be unattractive as well as cause aisle height compromises by lowering the ceiling height in the places where they cross the aisle. Low ceilings can be a problem should a horse "act up" and rear. I would suggest that you try and avoid running utilities across the main barn aisle or limit any such need to only cross over short aisles such as the connector aisle between the barn and the arena.

NOTE: All heated rooms should be constructed with 2″ x 4″ lumber in order for the walls to be thick enough for proper insulation. Spray in, foam insulation gives a great, tight, insulating seal, and can be sprayed into the walls in a quantity that yields a large insulating "R" factor and a very tight barrier against air circulation gaps that cause heat loss. I recommend that you use spray in, foam insulation for all the exterior walls of every heated room in the barn. The cost of low density, spray in foam is just a little more than fiberglass batts and the results are much better.

The following is the barn floor plan I finally drew up that has worked very well for me. I offer my barn plan to give you a visual idea of what we have been talking about, and to use as a starting point for your own barn plan.

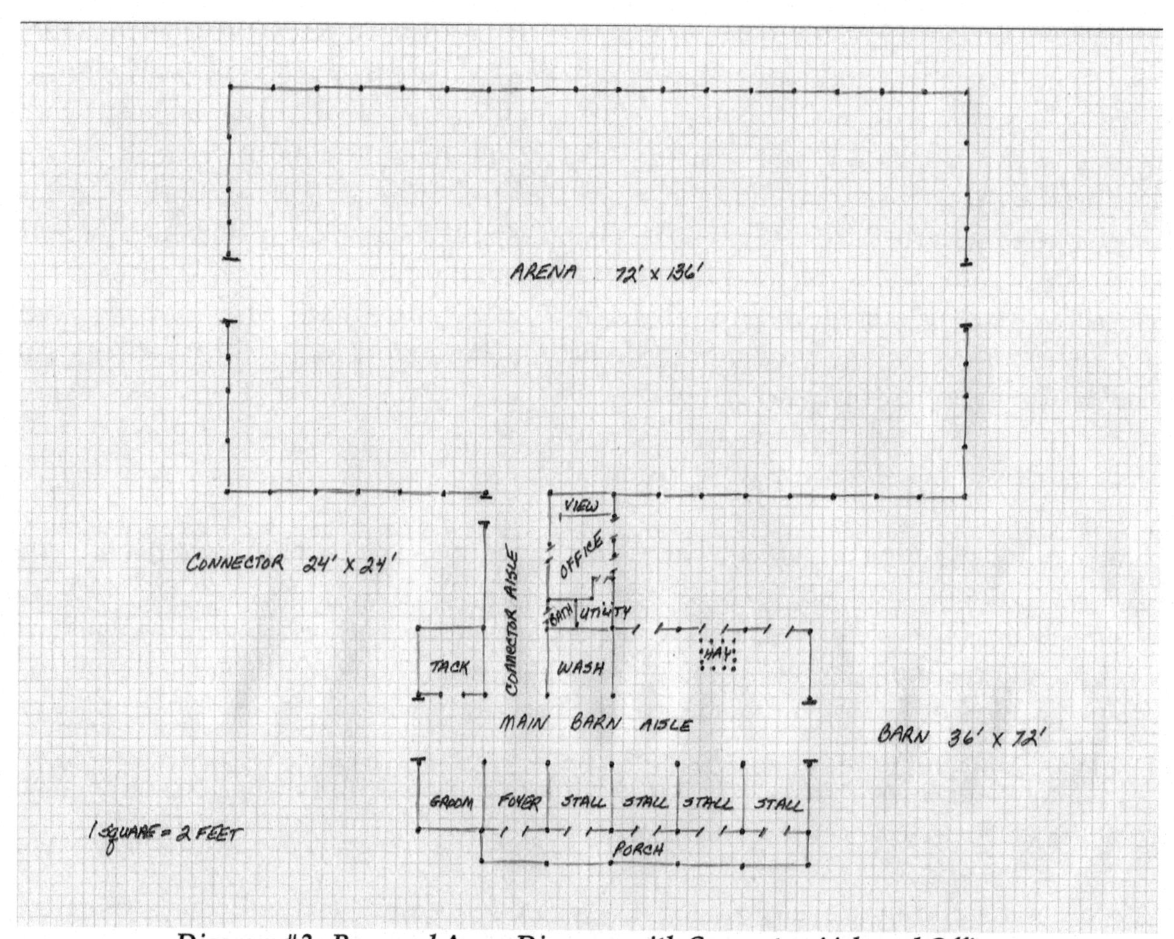

Diagram #3: *Barn and Arena Diagram with Connector Aisle and Office*

NOTE: One of the things to note about my barn floor plan is that it can be easily and inexpensively expanded—both the barn and the arena—which I did to be able to accommodate my planned commercial facility and future expansion. Having such flexibility is also good for the private facility should you decide to acquire and house more horses than during your original planning.

Now I'll talk about each of the areas in my barn plan and explain my thinking behind each area.

We've already talked about how big the stalls should be and why. Now let's talk about the rest of the stall.

First you should consider whether or not you want to have **automatic waterers** in your stalls. You need to take this into consideration now before you put in your stall floor footing because pipes and electrical wiring for heaters in the waterers will need to be installed under the stall floor, below the frost line, if you want to have automatic waterers. The best automatic waterers have an option for meters to measure the consumption of water by the horse. You've probably noticed that very few horse barns actually have automatic waterers, and yet, water is a huge issue with horses. Horses must have continual access to a fresh (and clean) supply of water.

You can of course address the issue of water for your horses by simply hanging water buckets in the stalls and using a hose fed from your freeze proof water hydrants to fill the buckets. But remember that those buckets will freeze in the winter in northern climates unless they have heaters and will need to be refilled on a regular basis, a couple of times a day at a minimum, and either you or a worker has to make sure that it is done. Also, remember that if you are filling buckets from a hose, you will have to drag out the hose every time you need to fill the buckets and then recoil the hose when you are done. You will also have your horses shuddering at the back of their stalls every time you are refilling their buckets as they anticipate that this might be the time when that sudden, out of control, water jet, that so often occurs while you are attempting to access the water bucket with the scrunched up hose, will happen and zap them. Don't forget that you will also need to scrub out each bucket everyday. Just filling the hanging water buckets several times a day is a time consuming and labor demanding task which will become overwhelming to you over time.

NOTE: Be sure to plan where you want your **freeze proof hydrants** to be located inside your barn (as well as any you want installed outside the barn for outside watering). In northern climates, you will need these freeze proof hydrants regardless of whether or not you choose to have automatic waterers.

Water was a constant point of caustic contention between me and the owner of the facility where I boarded our horses. I just could not come to terms with being told that I should not fill up my horse's empty water bucket on a 95 degree day in the summer. Worse, I just cringed when the facility owner would warn me that I had "better not get my horses used to having water." What in the world was that supposed to mean?

Then, this same judgmental owner-trainer would suddenly flip her logic when we were at horse shows and start yelling at all of us for not making sure that our horses' water buckets were constantly filled up to the very top. Talk about a double standard. Can you tell me how come the same horses who were deprived of water at home and were not allowed to "get used to having water" were now at a horse show and suddenly in need of having constantly full water buckets and were allowed to get as "used to having water" as they wanted to?

Photo #31: Horse Barn Aisle Freeze Proof Hydrant

Then there were the ice topped buckets hanging in the stalls in the winter in the boarding facility from which the horses had no interest in trying to drink the water. The facility owner's reasoning here was simply that "horses don't drink as much water in the winter," and, therefore, I was not to complain about the ice. However, I couldn't help but notice that when I added hot water to the ice covered bucket and made warm water available, that my horses would immediately drink down the entire bucket of water. And why was it that so many of the horses were getting impaction colic in that barn in the winter? Wouldn't impaction colic just kind of logically scream to you that it is a result of the horses not having enough water?

Photo #32: Outdoor Freeze Proof Hydrant

Well, folks, here's the truth of the situation that I found out both through research, before I actually installed the automatic waterers, and the reality of observing my horses who now have a constant supply of fresh water. First, the truth about why the horses were being deprived of water had nothing to do with them being unable to handle getting "used to having water." It was more simplistically that the more a horse drinks, the more he urinates. The more the horse urinates, the quicker the shavings get wet and "used up." The quicker the shavings get "used up," the more often you have to change the shavings. The more often you have to change the shavings, the more it costs you for shavings per horse. The more you have to pay for shavings per horse, the less money you make in profit off your boarders. That's all folks!

It's just about money. It had nothing to do with the physiology of horses. It simply had to do with the owner's bottom line profits. Therefore, the double standard between being home and at shows was finally revealed! It suddenly all made sense. Meanwhile, I had to live with the sickening reality that for years my horses were in fact being deprived of water.

NOTE: I was able to do my "water consumption" research because in addition to installing the automatic waterers, I also installed "consumption meters" on each stall. Consumption meters provide a great "heads up" for letting you know if your horses are drinking enough water. **(see Photo #88)**

The second issue of horses needing less water in the winter was also not completely honest either which I came to discover as I kept track of the water consumption meters through the seasons. Yes, horses don't drink the kind of excessive amounts of water in the winter that they might drink on a beastly hot summer day, but they still require a basic amount of water for healthy bodily functions from a steady supply of fresh water. The real truth of why the horses weren't drinking from those ice covered buckets is that horses will not drink water that is below 45 degrees or above 70 degrees. They want and need to drink water, but they can't palate water that is colder than 45 degrees. This is especially true of older horses with sensitive teeth.

What is so wonderful about our excellent automatic waterers is that there are water heaters in them that are activated in the winter season and keep the water from ever going below 55 degrees. Then in the summer, the water is always cool because it fills on demand from the naturally cool well water. Therefore, your horse will drink all day long in the winter the same basic amount he will drink in the summer.

In general, a horse that weighs between 800 and 1,000 pounds needs a minimum of 8 to 10 gallons of water per day in winter and summer or about 1 gallon of water per 100 pounds of body weight. In summer, the water need can rise to as much as 18 to 24 gallons per day. If you don't have automatic waterers, then you must refill buckets several times a day in summer and/or hang at least two buckets per stall.

Note: For the commercial facility, like the one I used to board in, having the horses drinking more water in the summer then requires that the pails be filled up more often which then raises the issue of having to pay someone to fill them up more often. Therein lay another reason why our boarded horses were not allowed to "get used to having water" in the summer. Again, it was just all about money while the horses weren't getting enough water.

So rather obviously, I'm encouraging you to install automatic waterers (specifically, counter weight balanced, auto refill ones) as well as usage monitors (water consumption meters) on each stall front. You will be so happy that you did. You will never have to worry about your horses having fresh water available to them whenever they need or want water. You will also not have to be constantly vigilant over the water consumption of each horse and refilling each bucket or upset to find that a horse has been without water.

But you do have to remember to turn on the water heaters when the weather turns cold. I can't tell you how shocked I was the day I came into the barn on the first late fall day on our farm when the outside temperatures had dropped into the teens. To my total horror, I found that the water in the waterers had frozen into an ice ball in each of the horse's water bowls, which had in turn resulted in the horses being in a mild state of panic! Of course, in our instance, I also found on that day that our charming "first" builder had failed to install a switch anywhere to enable us to turn on the heaters in the automatic waterers! Nor, for that matter, did he have switches to turn the pasture waterer heaters on OR off!

NOTE: Be sure the heaters for each of the automatic waterers are installed on separate switched circuits so that you can turn off any one of the waterers at any time. And I would also advise you to have separate circuits for the inside and outside waterers for the times when you don't plan to use the outside waterers in the winter and want to turn off the heaters.

Remember that the flipside of turning "on" the heaters in the automatic waterers inside the barn is true for the automatic waterers in your pastures. If you aren't going to be regularly turning out your horses in the winter or you aren't going to be turning them out for any extended period of time, you must remember to "disconnect" or switch off the heaters in the outside waterers or you will have a gigantic electric bill. Of course without our builder having installed a switch to turn off the heaters in our pasture waterers, this is exactly what I had—a breathtakingly huge electric bill! Please note that you should also disengage the automatic refilling feature in pasture waterers if you disconnect the heater for the winter.

There are also water bowl covers for the best automatic waterers when not in use. By using covers on your pasture waterers, you will keep out polluting bird droppings, insects and other debris from accumulating in unused waterers.

Also remember that nothing is perfect—especially things mechanical in nature—and there will be times when the water doesn't shut off when refilling in the automatic waterer and the bowl will overflow which will require an adjustment to the counterweight to correct. Then there will be other waterer glitches that arise here and there, but overall things go well. The water bowls in our waterers are also very easy to pull out and clean every day—just dump, wipe, rinse, and return to the stand. In order to feel comfortable about dealing with the waterers, be sure that you get all the repair and maintenance information from the supplier, and also "visit" a dealer at an "equine show" and have them fully describe repair and maintenance of the waterers to you. Also get the phone number of the manufacturer to call when you can't figure out for yourself how to repair something and/or how to get a replacement part.

UPDATE: As I mentioned, things do go wrong with automatic waterers. Here's what happened the other day just to give you an example of potential drawbacks.

My husband and I went over to the barn and entered through the office door. As usual, he stayed in the office to feed the cats, et al, while I proceeded into the barn to feed the horses. When I opened the aisle door from the office and walked out into the barn aisle, I was stunned to be greeted by a river of water flowing down the middle of the main barn aisle and out of the barn end door. Quickly trying to size things up, I ran down the connector aisle, then through the river and over to the first stall from which the source of the "head waters" seemed to be emanating. What I first found was Ian—our 16.2 hand Thoroughbred—standing in the far corner of his stall, shaking with fear. My focus next immediately turned from Ian to his automatic waterer which was pumping out water at an alarming rate. My eyes next dropped down to see Ian's entire stall floor flooded with more than 3 inches of standing water (proving that our stall mats—which I'll talk about a little later--are in fact impervious!), with the excess water flowing out into the aisle, then down the aisle, and out the end door of the barn.

I quickly entered Ian's stall to shut off the waterer and stop the flow of the river at its source, but found that when I engaged the water shut off mechanism, the water didn't shut off. No, in fact, to my total horror, the water just kept gushing out of the waterer. At this point, I left Ian and ran back to the office and told my husband I needed his help really quick! I then ran back to Ian's stall which I now had to enter and then proceed to convince one very large, frightened Thoroughbred that he should follow me, walking first through the stall floor lake of water, then past the scary waterer, across the flowing river, and down the aisle to the arena. My mind quickly calculated that my odds were really low of getting Ian to come out of the stall without a lot of trouble, including such things as him outright balking or backing, and even perhaps rearing and striking, which he has been known to do. However, happily, and rather amazingly, my "no nonsense" attitude, which is also known as the basically threatening "follow me or else" attitude, inspired Ian—at least in this instance—to follow me completely compliantly! I have to admit that I also suspect that Ian was desperately anxious to exit his scary stall as soon as possible and instinctively chose to be as cooperative as possible with anything or anyone who had the "balls" to volunteer to be the leader at that moment!

Meanwhile my husband knew where the main shut off valve was inside the waterer and was able to stop the gushing water once I had Ian out of his stall. As a first step to correcting the water disaster, my husband and I began sweeping all the water out the end door of the barn. But then, of course, because the outdoor temperature was a mere 8 degrees, and the indoor temperature of the barn was a correspondingly low 15 degrees, as we swept the aisle water and the water got shallow in the aisle, the shallow, aisle water immediately turned to ice. Then because the aisle was all ice, we began slipping all over the place as we tried to clear out the rest of the aisle water. So with the ice forming, and us slipping, we now had to stop sweeping out the water and quickly go get road salt from the house garage and spread it around the barn aisle so we wouldn't slip and break our necks. We also spread the salt all along the end door track of the barn doors so they wouldn't freeze open—good thinking there! But, meanwhile, of course, the thing we forgot in our panic was that we could have turned on the in-floor heating we had installed in the barn aisles and that would have melted the ice.

The aisle water now under control, we next had the lake in the stall with which to deal, and that of course contained floating feces as well as urine soaked shavings. I went right to work mucking out the HEAVY water slogged mess into the wheelbarrow while my husband got the wet/dry vacuum and started sucking up all the water he could. Meanwhile, we had a business meeting appointment for 1 p.m. in a city located about a one and one half hour drive from our barn, and it was 8:30 a.m. when we first found the lake and the river. Zoicks! With the appointment in mind, we worked quickly and got things "under control" and were able to return Ian to a nice, dry stall with a water bucket hanging on it. Meanwhile we raced to take quick showers and then drove like maniacs up north to the business meeting.

When we finally located a local repairman to come fix the waterer, he discovered that the flood had been caused by the heating element inside the waterer melting through the rubber water hose making a huge hole in the hose. The huge hole in the hose then caused the consequent gushing water problem and my inability to turn off the flow with the counterbalance shut off control.

Just another day at the farm!

NOTE: When we called the waterer manufacturer, they claimed they never heard of such a thing happening and felt our waterers must have been incorrectly installed. Well, of course they would say that! In the meantime, what we've done is to just put extra rubber hose around the actual water hose as a "guard" and then we check the "guard" hose for any heat damage on a regular basis! Point of note is that we had their "authorized installer" install the automatic waterers in our barn!

However, since this has happened, it brings up the need for installing an "alarm" on your waterers to let you know when an overflow problem is occurring. You should also be sure to have individual circuits on each waterer in order to be able to shut off each waterer's heating element. Even with this catastrophic failure "event," I still vote for automatic waterers! Just a couple of days of "making sure" Ian's water bucket was filled with warm water while his automatic waterer was broken, and having to manually fill a bucket several times a day, quickly reminded me of just how convenient the automatic waterers really are!

I should also share with you that there is some "getting used to" for the horses in regard to their automatic waterers. The waterers are decidedly "different," and they do make noise, and they do have movement! I'll talk about how my horses got "used to" their waterers later on.

NOTE: If you don't want to install automatic waterers with heaters, you should still consider providing your horse with warm water during the winter months. You can supply warm water by using water buckets with preinstalled embedded heaters which are plastic buckets with a built in heater in a false base plus an electrical cord to plug into an electrical outlet. These buckets also have thermostats to control the temperature. If you choose this method, you must remember that you will need to have an electrical outlet installed on the aisle side of each stall. Providing your horse with a constant supply of fresh, temperature appropriate water is one of the single most important things you can do for your horse's overall health.

Now, let's talk about the next important decision you will need to make about your stalls which is the composition of the **stall floor**.

Your basic stall floor choices in northern climates are clay, packed gravel under rubber matting or cement. In southern climates, hard packed sand mixed with gravel is an option.

Our horses lived in clay based stalls for years. I don't have anything particularly good to say about them. First of all clay is essentially non porous and as such, urine does not percolate freely down into the soil under the stall which makes the stall essentially dirty. If there is any percolation at all, it's painfully slow at best. In addition, clay is cold and damp. Maybe the cool dampness is refreshing in a limited way in the summer, but it's miserable in the winter especially for the aging or over worked horse with arthritis. The way the clay floor system works is that a "catch basin" depression is made in the center of the stall. Shavings are then put over the catch basin area to absorb the urine.

Theoretically that sounds okay, but the first thing I noted over all the years is that there was a tendency on the part of the barn owners to keep shoveling the perimeter shavings into the center basin over the several days before a stall would "qualify" for new shavings. That meant that by day four since new shavings were put in the stall that the stall floor had become nothing but a clay bowl with no shavings covering the perimeter clay and, therefore, the perimeter of the depression did not offer the horse anything warm or soft to stand or lay on for about two out of every four days.

But worse, the combination of the small stall size at the boarding facility, and the pitch of the overall stall to the depression in the center of the stall, meant that the horses had no real flat area on which to stand while in their stalls. I always wondered why the horses were so in love with pressing themselves up against the dividing walls between the stalls. I thought maybe there was this overwhelming need of horses for companionship, even when in their stalls, because the horses were so consistently wedged against either one side or the other of their stall walls. It wasn't until I was building our farm that I realized that what the horses were merely trying to do at that facility with the uneven clay floors was to find the flattest available piece of stall flooring on which to stand so their footing would be even and not stress their ligaments and tendons. Realizing that the uneven clay floor was continually stressing their ligaments and tendons then explained the other perpetual issue at that facility of why all the horses had continual, intermittent episodes of "mystery" lameness.

The next option for stall floors is cement. Our state university veterinary center has these, as do a lot of veterinary facilities. But the key here is to remember that horses don't stay long at any of those facilities, whereas your barn will be your horse's house where he lives and stands day and night, day after day. Therefore, the footing in the stall in your barn must be maximally comfortable for your horse and that would not be cement. One thing cement stall flooring can be, however, is flat. It will be able to keep even the largest horse on level ground. Cement is also porous and therefore liquids CAN percolate down through them. However, that same porosity means that urine IS percolating down under the cement floor, and come the day you might want to clean out under the cemented stall floor, how are you going to do that? You aren't—without using anything short of a jack hammer. Of course, you can do a regular dilute bleach wash over the cement as they do every evening at veterinary facilities. However, you will quickly find that continually washing down the stall floors is impractical in that, unlike a veterinary facility, your horses don't leave and go back to another barn so that the whole barn is empty and easily washable. There is also the issue of bleach fumes which are bad for the horses to inhale, especially on any kind of regular basis. Plus you just won't have the time or energy to be frequently washing stall floors.

In addition, cement cracks and the ground under the cement can collapse with sink holes that are then hard to fill. If you think about any cement surface you have ever seen anywhere (except maybe in semi-tropical and tropical climates), you'll remember "seeing" cracks. Especially in climates where there are freezes and thaws, you can pretty much count on at least minimal cracking of cement regardless of how "solid" your base is. Cracks make for a big problem in a stall because they will allow just enough seepage of waste products into the cracks and under the cement to cause problems with mold, bacteria and fungus in the cracks and under the cement pad, while being totally inaccessible for clean up due to the cement covering.

There are spray-on sealants to waterproof cement floor surfaces. However, these sealants would most likely make the stall floor too slippery for horses. Therefore an additional covering over the waterproofing would be necessary, such as rubber stall mats. Even if you used impervious, rubber stall mats to cover the cement, you would most likely have run off of urine along the wall edges of the mats that would seep down to the cement base. Then you would need to frequently—if not daily—wash out each stall so that no standing liquids would be under the mats. That in turn would require you to have perfectly pitched cement aisles into a central drain system in the middle of your barn aisles that would allow the water to flow out of the stalls into the barn aisle and then out of your barn from the aisle drain—water that would of course freeze in the aisles in the winter unless your barn is heated.

I have also seen sealant covered, cement stall floors that have indoor/outdoor carpeting covering the surface to prevent the horses from slipping. However, this carpeting does require daily wash offs with water and a disinfectant such as bleach and would require the same kind of costly central drainage system in your barn aisle as mentioned above. Well, not only costly, but trying to get a cement contractor to "perfectly" angle anything is pretty close to an impossible task for the average human being to accomplish!

NOTE: Whereas I am not an advocate of cementing stall floors, I do like cement for the barn aisle floors! Be sure, however, that any cement floors are given a **brush finish** to make the cement less slippery—do **not** have a smooth finish on cement barn floors.

Also, like clay, cement is cold to stand on full time and would be hard on the aged, overworked, and arthritic horses. You could however, not only put rubber stall mats over your cement floor but also add shavings on top of the rubber mats. Nevertheless, if there is urine run off from the mats, down under the mats to the cement floor, you will now have the double problem and expense of having to remove both shavings and then the mats to clean under the rubber mats on a regular basis (and the correct thickness rubber mat you would need to use is VERY heavy and awkward to move around!). You would also still have the original problem of the run off percolating over the edges, down through, and then under, the cement floor where it can cause fungi and bacteria to grow, while remaining inaccessible to cleaning if the cement is unsealed.

The other option for a stall floor is to have several inches of compacted limestone dust placed on top of several feet of compacted sand, covered with 3/4" rubber mats, and topped with shavings. At first this system just sounded "unclean" to me. It seemed like I was inviting a giant urine retention pond to occur under each stall. However, when I thought more thoroughly about this system, I came to realize that this method was actually the equivalent system to a basic "human waste" septic system being based on natural percolation of liquids down through sand and soil.

The traditional way for this stall base to be formed is to remove several feet of natural substrate soil. Then several feet of sand is added and compacted. The compacted sand layer is then covered by about 4 inches of stone dust which is also compacted. Then rubber stall mats and shavings are used above the footing.

We slightly modified the traditional base for this type of system and it has worked very well for us. We formed the base by first removing about 6 feet of the natural substrate soil from each stall area (this also allows for the automatic waterer lines to be laid in the stalls below the frost level). Then masonry sand was put into that hole in each stall area to fill in to an approximate depth of 5 feet in the hole. The masonry sand was then covered by about 5 inches of 21AA stone. The 21AA stone was then covered with about a 4 inch layer of stone dust and compacted. After the stone dust was compacted, the stall base was topped with rubber stall mats. The stall mats were then covered with wood shavings to absorb the majority (if not all) of the urine. The shavings need to be cleaned, picked, and partially replaced at least once every day.

The stone dust gravel used in this stall system is of a composition that quickly congeals into a cement-like strength with a minimal addition of water. As such, the congealed stone dust forms a very firm, "cement like" base. The congealed stone dust, however, is also porous and allows for percolation of urine down through the stone dust to the stone layer and then to the sand, where the sand further percolates the waste run off, and cleans it, so that by the time the liquid waste is returned to the underground water system, it is environmentally safe and clean. In addition, the stone dust, unlike cement, can be relatively easily dug up and removed or replaced when needed.

NOTE: Some experts feel that the gravel base should be replaced every year if you are using non-impervious, non-interlocking, rubber stall mats because there will be run off on the edges and through the seams of regular rubber stall mats. However, I have found that impervious, interlocking, rubber stall mats only require base redo every couple of years because there is, at most, minimal, if any, urine seepage below the mats.

I chose this system for my stalls and felt very happy with the solidness of the flooring and the mini septic system concept. That was until my farrier told me that my horses had thrush and that it was probably due to fungus and bacteria growing under my rubber stall mats which were ultimately migrating up into my horses' hooves. I was upset!

The first practical issue in front of me was that picking up the rubber mats and cleaning under them was just out of the question. The weight of the mats alone in one stall would just be too much for me to handle. So all I could do to address this accusation by the farrier was to religiously clean the top of my mats with dilute Lysol and dilute bleach to try and kill off any fungus or bacteria that might be growing on top of the mats. I went obsessive over the whole thing and agonized endlessly over the situation.

Then came the day when the stalls "collapsed" from the back up of the down spout drainage and my husband and I HAD to pick up the mats and add stone dust to the stall base floors as I discussed earlier. (I should point out here for any potential cement stall floor enthusiasts how impossible it would have been to do anything about the collapsed stall floors if the downspouts had caused the back flow and undermined the footing under cemented stall floors!) When we rolled up the mat sections to repair the sunken stall floors, we found that the underneath of the mats was totally dry and mold free. Absolutely nothing had been percolating "down" under the rubber stall mats from above. Our only problem that day was the percolating "up" of the backed up downspout water into the stone dust from below, which in turn made the stone dust congeal even more solidly than the builder had tamped it to be in the first place, and, therefore, collapsed the stall floor. Under-mat fungus and bacteria were not an issue.

Two lessons came out of this situation.

LESSON #50: First, that you shouldn't necessarily believe everything your farrier tells you, because, in fact, it turned out that my horses didn't even have thrush!

LESSON #51: Second, that the type of stall mat you choose is critically important. Interlocking mats are the best.

The farrier could have been right about fungus and bacteria growing under the mats if we had used "ordinary" rubber stall mats. Ordinary rubber mats are ¾ inch thick, however, they are made from a porous type of rubber and they are cut with straight edges which you have to try and align together as closely as possible to minimize seepage. Because we are human and the natural state of things for humans is far from continual perfection, getting straight edged mats to perfectly align is probably not going to happen. Or if it does happen at first, it won't last as the weight and movement of the horses cause shifting of the mats, and the porous property of the rubber in the mats causes shrinking. In other words, sooner or later there will be gaps between your straight edge rubber mats, and those gaps will most likely be sufficiently large enough for a significant amount of urine et al to actually seep down and spread under your stall mats, potentially causing problems. In that case, you would indeed have to pick up and clean under those mats on a wearisomely regular basis.

But if you buy **impervious rubber mats**, with **interlocking edges**, you will have reduced the amount of percolation down under the mat to the point of being so minimal as to be pointless, and therefore require you to do little, to no, "under mat" maintenance. So I recommend you purchase these "humane," impervious, ¾" rubber mats with the interlocking edges. Our horses LOVE standing on them—the horses are sound—the mats keep the horses warm in the winter and cool in the summer—and the horses are happy.

Of course there are also **shavings** covering the rubber of the mats to basically absorb all of the urine. I keep a good one to two inch covering of wood shavings over the entire mat area. Once your horse has determined his regular urination area, you can make a point of keeping extra shavings piled in that area to catch the urine. Also—and this is very important—you should put down a granular, non-toxic, horse friendly desiccant (drying agent) over the frequently urinated area. Using a desiccant over the rubber mat and under the shavings, concentrates the urine and keeps the urine from spreading through, and unnecessarily wetting, surrounding shavings. It will save you a lot of money in shavings and a lot of time in mucking if you use a desiccant.

To save money on shavings, some people only put shavings on the area where their horses frequently urinate. However, I cannot for the life of me figure out how this works. Yes, horses do have an area where they most frequently urinate—remember, however, this is mostly true only if the stall is the correct size (large enough). Yet, horses don't ALWAYS urinate in the same area regardless of any factor! So there will be days when the horse doesn't "hit the spot." Without shavings evenly spread over the stall floor, there will then be "open" pools of urine here and there on the mats. Plus there will be piles of horse poop sitting on top of bare mats. Without any shavings, the poop piles won't be distributed through or "lost" under the shavings. That means the droppings remain fully exposed, and concentrated. And, that means when your horse walks over these piles or lays down on them, he's for sure going to be both covered in feces and have feces wadded up into his hooves. Plus, even if your horse always "hits the spot"—which is unrealistic—the shavings on the spot aren't going to stay in one spot. The horse is going to walk over and through that "spot" of shavings and gradually spread them around the stall. So ultimately, "the spot" might not have any shavings on it at all by the time the horse tries to "hit" it!

As things currently stand with my having two gray horses—who are at the white-haired stage of their lives—even with shavings totally covering their entire stall floor, when they lay down in their shavings-distributed poop, and then stand up, I suddenly have "paints" for horses rather than grays! Can you imagine how my gray horses would look after laying down on stall mats with just the one spot of shavings approach? Actually, I don't want to imagine.

Meanwhile, I just can't understand how people manage this system. My only conclusion is that it works for bays and other dark coated horses and for them alone. However, if you consider the "filth factor" and the reality of what the dark hair is covering up on the dark colored horses, then the visual of the reality of the poop and urine stains that are there, but hidden on your bay, brown, chestnut, and black horses, could make you feel kind of sick.

Okay, so let's move along.

Let's talk about **stall separator walls**—meaning the walls between each individual stall that separate the individual horses.

At first, I wanted the stall walls in my barn to be open. What I mean by that is I wanted to have just bar grills between each stall above a minimum, four foot high, kick wall, so that: (a) the horses could totally see each other and have the security and comfort of knowing where the herd was, and the status of all the members of the herd, at all times; and (b) the air could flow freely down the aisles and through all the stalls. Plus bar grill stall walls are pretty for us humans to look at.

However, as much as I wanted them, I couldn't get anyone to agree with me that bar grills were the correct choice for stall walls. I couldn't even get anyone to agree with me that there should be a window-sized, bar grill "insert" in the otherwise solid wood stall walls everyone was insisting I have so that I could at least give the horses the option to see each other and allow limited breezes to blow down the aisle and through their stalls.

NOTE: Bar grill, stall separator walls must be purchased from a supplier. Your builder generally will NOT build such a wall for your barn.

I couldn't get any sympathy for the aesthetics and horse friendly reasons I thought bar grill, separator walls would provide. Rather, I was told that sooner or later one horse would bite the neighboring horse's tongue off or do something even worse (although I would vote that having your tongue bitten off is "worse" enough), and therefore I could not have what I wanted and that was that.

I guess you can tell that I wasn't happy about being forced to give up on my bar grill, stall separator walls. From that point of rejection on, I found myself just cringing every time I was required to envision how the final barn would look with #2 southern pine, tongue and groove, solid wood walls going up between each and every stall, rather than the beautiful bar grills I so wanted.

The first thing I have to confess to you now that I actually have the solid wood walls separating the stalls, is that the tongue and groove pine actually is pretty to look at and gives a very warm look and solid feel to the stall as well as to the barn. That surprised me. The second thing I'll confess is that I do believe now that I have the solid walls, that the horses seem to actually prefer the total privacy of their stalls. They clearly feel peaceful, relaxed, safe and unthreatened in their stalls.

However, the one thing that remains a negative about the solid wood walls in the stalls, and I remain unhappy about, is the air flow situation. Even though our barn has plenty of air flow with a window in every stall, etc., there are still those really hot summer days when I turn on the fans in the aisle and I wish the air could flow freely and equally down and through all the stalls.

One thing that I would NOT compromise about, however, and is the thing I believe makes the solid wood separating walls look okay, is the **stall fronts** I selected. Here again, I was warned that I should not even consider having the low profile, European stall fronts. I was told that there would be constant fighting as aisle horses passed by the horses standing in their stalls. Well, I just didn't feel that was going to be that big of an issue, and so I chose for the European stall fronts regardless of anyone else's opinion. But, the truth is that, depending on the particular horses you have, and the arrangement of your stalls along the aisle, and the width of your aisles there could be "passing" horse issues.

NOTE: Generally speaking your builder is not going to "build" **stall fronts** for you. Stall fronts are specialized, individually purchased units. You should look online for suppliers and styles and prices to determine what kind of stall fronts you want. The stall fronts are shipped to your barn site and then the builder will install them. Be sure to know exactly what size stall fronts you are going to have so that the builder correctly allows for the width of the stall fronts when constructing the stall separator walls. The best plan is to have the manufacturer send you all the specifications for the dimensions and installation of the stall fronts you have selected, and then give **ONLY** a **COPY of the specifications** to your builder. Keep the original in YOUR file at home! Rest assured, the builder will repeatedly lose his copy of the specifications and be asking you for another copy!

NOTE: Each pole barn supplier does have its standard, basic, stall front door available for you to purchase. Their standard door will most likely be substandard to those manufactured by stall front manufacturers in my opinion. Remember that stall fronts are more than about the wood. The safe finish of all edges, and the quality of all parts of the stall fronts must be considered from hinges to slider tracks if, for example, you should choose slider doors.

When I took a riding clinic in England, the horses for the clinic were stalled in a small, rather temporary type of barn structure. It also had a trailer structure used as an office area on the outside wall of the barn where the roof extended out from the barn. Inside the barn were the type of temporary vinyl stalls you see under the canvas tents at big "A" shows with stall fronts that allowed the horses to stick their heads out into the aisle (like the low profile European stall fronts I selected). There were problems with the stall fronts.

The problems that arose weren't so much that the horses could fully stick their heads out of their stalls, but that the aisle was too narrow. How narrow was the aisle? Well, the aisle was so narrow, in fact, that the horses could just about reach across the aisle and bite each other. So naturally, you can imagine that we humans walking down the center of the aisle had little opportunity to avoid being perfect bombing practice targets for each and every horse who might be inclined to take a snarly swipe at us. There was absolutely NO avoiding the horses who wanted to nip at us. I think you can understand that the situation, shall we say, was VERY uncomfortable at best, and why I came to hate walking down that frickin' aisle and ducking around and under attacking horses. I eventually found it much happier to consistently take the exterior route to enter the barn using the alternate short aisle and avoid walking through the flack zone even though the exterior route made it longer and harder to carry tack and equipment from the other end of the barn to my horse's stall.

But none of the preceding story is a problem in a barn with a **wide enough aisle**. Remember, we are planning a barn for you in 12 foot sections, and that your barn aisle will therefore be a nice, wide 12 feet.

Some of the benefits of the European stall fronts are that when your horses want to socialize, or catch the aisle breezes, they can choose to do so by sticking their heads out over the low profile stall front and have both an aisle breeze and a full view up and down the entire aisle. These stall fronts are just wonderful for the horses and they make for very happy horses. Plus the European stall fronts are just flat out pretty! Everyone who comes into the barn LOVES them! It just sets a beautiful image for the barn. There are also lots of other little benefits to having these stall fronts like being able to put on or take off halters without having to open up the stall. And of course, it's very easy to give treats and kisses! It also freely allows your horse to nuzzle the back of your head as you stand in front of his or her stall and talk to one of those "stinky" human friends of yours.

Photo #33: European Stall Fronts with Full Bar Grill, Swing Doors

I was also told that it was a HUGE mistake to have **stall doors** that swing open. I was warned to ONLY have sliding doors. Well, that didn't make any sense to me. What in the world would be so inherently dangerous about a swinging stall door? What I was told was that the horse would get "caught" in the door, and all sorts of awful things would result.

The fact is, from my personal experience, that it's the sliding stall doors that are a complete nightmare. It seems to be more consistently inherent in sliders than any other door type, that a thing that slides will in time necessarily get stuck—and/or that a thing that slides will ultimately fall off.

Have you ever had the "wonderful" stall slider door experience of tugging on the slider door on a day when the humidity has swollen the wood header board, and made the door stick on the track, such that the entire, heavy wood, slider door comes falling off the stall front? I have. Let me tell you, it's not a good thing to have happen. Take a second to envision this—the heavy wood slider door suddenly crashing down to the floor ... your reaction ... your horse's reaction ... every other horse and human in the barn's reaction ... and the now doorless stall in which is standing an unrestrained, frightened horse ... standing there for the briefest of moments that is!

Nevertheless, my "advisors" had me nervous about the European stall fronts being low profile, and about the swinging stall doors. Although I kept trying to envision how things could go wrong with the European stall fronts, I just couldn't come up with anything short of the hinges failing and the door coming to fall off. But how would that crisis be any different from the stuck slider doors I have known?

Well, anyway, the bottom line is that now that I actually have this type of low profile stall front with its swinging door, I can say without reservation that I LOVE the low profile stall front as well as the swinging stall doors, and that, in my opinion, they are totally functional, practical, and convenient.

NOTE: I also got the full bar grills on the swing doors of my European stall fronts. I did this for air flow into and out of the stalls. Even though I do love these full bar grill doors, I have to warn you that if you have a horse that kicks at his stall door, that the powder coated finish (the colored paint over the metal) will chip off the metal bars and allow in rust. However, they can be repainted.

If you are running a commercial facility, however, I would feel the need to somewhat modify my enthusiasm and strongly recommend that you consider having one side of the aisle with the low profile stall fronts and then the other side of the aisle with coordinating full height stall fronts. I would do this just so there wouldn't be any problem in case you should happen to take in boarders with horses who aren't the friendliest of four footed beasties! I still wouldn't have sliders though if I could avoid it. I would get full height, swing open, stall doors which, by the way, are available.

The whole point here is that there are two basic issues with the stalls. One is horse safety, and the other is horse comfort. You want the horses to have as much visual ability as safely possible, and as much air flow as possible, and yet you want to provide them with a strong basis of privacy and security. All products have some inherently "evil" factor in them under the "right" circumstances that will cause them to become potentially dangerous. But by exercising care with anything you use, and always being vigilant for the potential of a dangerous situation, you should pretty much be okay with any of the products you choose to use, in my opinion.

One overall thing that we found was that the look, quality and serviceability of the barn is significantly increased by using quality, non-standard items (meaning items NOT from the pole barn people--such as exterior windows, doors, and stall fronts). In other words, you do NOT have to use the pole barn builders' doors, windows or stall fronts. You can be eclectic about your structure and get much more for your money, and much more for both your horses' and your level of happiness, by purchasing from several equestrian product suppliers, and then having your barn-finisher builder supervise the inclusion of those products into the otherwise standard pole barn structure. In other words, you may only be able to afford a basic metal barn structure, but by investing in quality windows and doors and stall fronts, for example, your finished barn will create an impression that will be one of a much more expensive barn.

We dealt with a wonderful supplier for the stall windows and for the barn's front door. I say "wonderful" because not only were the products of superior quality, but the company was more than willing to work with me to produce whatever I wanted.

What do I mean by that?

Well, let's talk about the **stall windows**. First of all, you should definitely plan to have a window in every stall. After boarding our horses for years at a windowless facility and worrying incessantly about how the horses were going to survive the stifling heat of summer without a window to let in a breeze or survive the boredom of having no view to see the outside world, my having windows in each of the horse stalls was a top priority on the list of items to include in our barn.

One barn we visited that was owned by a friend of ours, had installed double hung windows in each stall of the type that would be used in a house. She then had window grill bars installed over the double hung windows to prevent the horses from breaking the windows. The windows made a pretty look both on the inside and outside of the barn. However, double hung windows don't allow for very much air flow in that no matter how big the glass area is, you only get half of that area when the window is actually open to allow in the air. Also, without safety glass in the panes, you could have a big problem should the horse in fact break the window. I've seen it happen where a horse has kicked hard enough on the grill bars to break the glass behind the grill and then have the broken chunks and shards of glass fall back into the stall and cut through the horse's legs. So, when I saw her stall windows, I felt this wasn't a great option for our barn. We felt the need to look further for other stall window options.

NOTE: The standard pole barn windows are small, slider windows. These slider windows, just like double hung windows, only open up half the window area when slid open. Therefore, these slider windows allow in a very limited air flow. It is only casement windows, and Dutch windows, that fully open.

The first stall window innovation we came across was at a recently built barn one builder had us go see because of the innovative windows the owner had installed in her barn. He told us with great pride that they were European windows and that the owner had found them online. We immediately went online and looked up the window company before visiting the farm. The windows looked just great online in that they were made with safety glass, fronted with iron grills, and the windows swung out against the exterior barn wall and as such fully opened.

Perfect! This was just what I had in mind for our barn. What could be cuter than seeing the horses' heads sticking out of the window on a warm summer day, catching the breezes, and taking in all the sights and the sounds of the farm?

As sure as we were that these would be "the" stall windows for us, we were not overly excited somehow at the way the windows looked when we arrived at the dressage barn and actually saw them in person. It was summer and the windows were open, but the beautiful image we thought we would see, just wasn't there. One of the first things we noticed was that from the exterior, there was an unfinished look in the way the windows were cut into the siding which I later realized was due to a lack of window trim. In addition, the window was smaller than we had envisioned and the opened window appeared thick and clunky as it lay back against the exterior siding.

The barn owner was very gracious and showed us around her facility. When we came to discussing the windows—about which she was very proud—she did mention one thing to us that she said was a big, unforeseen problem with the windows which she wanted to warn us about, and for which she still had no solution. She told us that when the windows were opened, and the horses stuck their heads out of the windows, that the horses had a tendency to eat the paint off the metal siding.

I thought she was kidding. Why would the horses do that? But then again, in thinking about horses, why wouldn't they do that? Regardless, the barn owner wasn't kidding. In fact, it was actually a huge issue, and one that was ruining the exterior siding on her barn. She said that she had tried everything she could think of to stop the behavior, and that her most recent attempt at discouraging such activity was to smear "bute" on the metal siding to get the horses to stop eating the metal. Well, the bute may have been a quick fix, but good luck to her next time she needs to get bute into her horses for health reasons!

Then and there I decided that maybe it wasn't so important for my horses to stick their heads out of their stall windows after all. Plus, in thinking about it, the horses would be open to danger, should a passerby be so inclined, if the horses were sticking their heads out the windows and I wasn't at the barn. Also, fully opened windows would make the barn unsafe. If a horse has enough room to stick his head out the window, then a human can enter the barn through that same window. And speaking of entering through the window, a large, open window means that any form of wildlife can also easily gain entry into the barn through the window including everything from four footed, furry ones to winged ones.

And then, while reviewing all these thoughts in my mind, it also suddenly occurred to me, that for all their innovation, the European windows on the dressage woman's barn, could only be either fully opened or fully closed. The protective bar grills were attached to the window, and were, therefore, "one with the window" and as such, the stall windows were necessarily either fully opened or fully closed. That meant if you opened the window for air, then the horse's head was able to stick outside the barn and the horse would be eating the siding. It was an all or nothing situation. In addition, it occurred to me that those affixed bars caused another problem. How in the world would I ever be able to fully clean the window behind those bars? I wouldn't!

Photo #34: Open Stall Window with Screen and Closed Bar Grill

Looking into other stall window options both online and at trade shows, it turned out that all the manufacturers either had the grills attached to the windows as a single unit or they had an opening window with a separate, permanently fixed stall bar grill that couldn't be opened. A window that opened separate from a permanently fixed bar grill would be cleanable (but only realistically in warm weather or in warm climates when you could step outside to clean the opened window). As such, the windows in winter could actually get dirty enough that the horses couldn't look out of them anyway, ruining the benefit of having windows. In addition, having permanently fixed grills meant you would never be able to give your horses the option of sticking their heads out of the window if you wanted to.

Photo #35: Closed Stall Window

Clearly—at least to my mind—it would make more sense to have both options available on stall windows. Why did the grill either have to be part of the window or separate from the window and fixed? Why couldn't each—the window and the grill—be opened independently of each other? In that way I could open the window and leave the grill in place. Or if I wanted, I could open both the window and the grill and let the horses stick their heads out of the barn. Also by keeping the grill independent of the window operation, I could open the grill and clean or repair the window in the winter from the inside should I need to do as such.

So, now, here's where the window supplier earned the right to be called "wonderful." Although no one had ever before asked for such a window system as I wanted, the supplier worked with me and provided me with a stall window that opened outwards, and a bar grill that opened inwards while making sure that every single edge and turn of the metal of both the window frame and the bar grill frame was horse friendly. The windows were also a nice, big 4' x 4'. They were actually based on the "tops" of Dutch door, stall doors, also known as Dutch windows.

Photo #36: Open Stall Window with Open Bar Grill and Screen

NOTE: The stall window needs to have a latch on the exterior siding of the barn to latch into when the stall window is opened. This will hold the window securely open. The latch must be mounted onto a square piece of wood, approximately

6″ x 6″, that has been attached to the metal siding of the barn. The metal siding alone will not be strong enough to support the latch plus the weight of the opened window.

Photo #37: Stall Window Exterior Latch on Wood Block

I have also installed custom made screens on the stall windows in each of the stalls. The screens are attached to the window opening from the exterior side of the space between the stall window grills and the stall window glass. I decided to put in screens to reduce the ability during the day time of flies and birds to easily access the barn through the opened windows. At night, in the summer, when the windows are opened and the end doors are closed, having the screens on the stall windows keeps out the mosquitoes and bats and any other type of night creature small enough to access a barn through the separations in the grills. The screens have a wood frame around them and are screwed into place. The screens can be easily unscrewed for removal.

Photo #38: Front Entry "Dutch Door" to Main Horse Barn Foyer

Having these wonderful stall windows, I next thought about how the front of the barn would look with these nice, big, stall windows when compared to the little, ordinary man **door entrance** we had so far planned into the barn. I could envision that by comparison the entry door would look small and junky. So, I tried to come up with something that would make a stronger look and statement for the entrance into the barn of equal strength to the look of the stall windows. It suddenly occurred to me that the obvious, logical compliment to the Dutch stall windows would be to have a coordinating horse stall, Dutch "door" to serve as the entrance into the barn which would then have exactly the same look and make up of the stall windows. And, that idea turned out in fact to be perfect. Perfect, except for one thing. Once again, I was the first one to ever ask the manufacturer to make a horse stall, Dutch door work as a people friendly, man door with a locking door knob. The manufacturer's first reaction to my request was that they weren't sure whether or not they could "figure out" how to install human door hardware on a horse door. But, once again, they worked with me and solved the problem that resulted in a beautiful, matching front door to our barn.

This is the kind of thinking that you want to do with your barn. Think of the things that you really want beyond the "standard" and then find a supplier who will work with you to solve your "problems" to get those "little things" you really want that will make all the difference in the look, feel, and serviceability of your finished project--and will do so at an affordable price.

NOTE: Having 6 inch, rough sewn cedar trim (or other 6 inch, wood trim) installed around your front entry door into the barn, around the office entry door, and around any casement windows used in construction (as in the office area exterior wall), adds an attractive finish to both the doors and casement windows as well as a look of strength to the surrounding, otherwise plain, metal walls.

I've already mentioned the advantages of having a dedicated **grooming stall**– the floors of which, by the way, should be covered with nice, thick, rubber mats just like you have in your regular stalls, and should be an area that is "stall ready" should you need to convert it into an additional stall. It's a good idea as well to keep both the grooming stall and the wash stall out of direct line view of each other, as well as out of direct line view into the arena. This is helpful once again to keep your horse calm, and his or her focus on you while in either of these service stalls. It's also best to keep both the grooming stall and wash stall from having outside walls. On really windy days, a lot of horses feel nervous in stalls with an outside wall rattling as you try and work with them. Also, an outside wall is cold and not appropriate for a wash rack.

Photo #39: Grooming Stall and Stall Ceiling Lights

I would also **not** recommend having a window on the wash stall. Having a window on the grooming stall would be your call. On the one hand, it could be distracting, but, on the other hand, you would want to have a window in the grooming stall if it became a future stall.

The **wash rack** should have a cement floor that has a very gradual pitch to a drain. I wanted my drain to be toward the back of the stall because at the facility where I boarded for so many years, the horses were ALWAYS afraid of the centrally placed drain they had. My horses would do anything they could to avoid stepping on that drain. As a result, they stood and moved in some rather interestingly contorted modes while being washed, and sometimes just would NOT move over past the wash stall drain or even enter the wash stall at times for that matter! Consequently, I felt that if I put the drain in the back of a 12′ x 12′ stall, I would have a better chance of my horses feeling comfortable in our barn's wash stall. But, of course, when I came over to see the progress on our barn construction, I found that the builder, who fully knew I wanted the drain placed in the back of the stall, chose to plop the drain right smack in the middle of the stall floor, and then had the floor cemented over. Ugh.

Surprisingly, however--and happily--I haven't had any issues with the drain here or with my previously "centrally-based-drain" fear ridden horses. Go figure. Actually, I'm quite sure the lack of fear over the central drain in our barn has to do with the spaciousness of the 12′ x 12′ wide wash stall. The horses seem to feel here that they have enough escape room should that drain monster actually ever pop its ugly head up through the drain grate on the floor!

In my travels to view other peoples' barns before building ours, I did see one barn with a wash rack where the cement floor was pitched forward into what is frequently called an "alligator" drain that fronted the entire width of the wash stall. This was a drain style for the stalls that our barn designer (Miss South) also favored. They are long rectangular drains of about 10 inches in width with lots of drainage "teeth" openings (which probably gave rise to the alligator name), and are generally installed in the floor across the entire front of the wash stall. However, I noted in the barn we toured that water was not only running into the alligator drain, but also running right over and past the alligator drain and out into the aisle. I don't know the particulars of why these drains weren't working effectively. I just know that they weren't working while I was there. I offer this information as a "heads up" for maybe being careful not to be too creative as to where you place the drains for your wash stalls.

There are wash rack specific, rubber, floor mats, however, I haven't purchased them and so I can't comment on how well they work. Two points I will mention are: (1) that my horses don't slip on the wet, "brushed" cement of the wash rack floor, so I don't necessarily feel a need to have rubber wash rack mats to prevent slipping; and (2) that I would think bacteria and fungi would happily grow under moist wash stall, rubber mats, and potentially become a breeding ground for the kinds of things you want to avoid exposing your horses to unless you want to regularly and vigilantly remove the mats and clean them—which is really a lot of seemingly unnecessary work. But, that's just speculation on my part and offered merely as food for thought.

I also saw ceiling mounted hose systems in wash racks that allow the hose to swing left and right over the top of the horse rather than having to move the hose around under the horse's hooves where the horse can potentially step on the water source and cut it off—which of course horses invariably do! Once again, I don't have any personal experience with using the above-head, water system, but it looked very awkward to me. I also felt that an overhead system would definitely be something that horses would need a lot of "getting used to" before you could work comfortably under it with them. It also looked like a maintenance and cleaning nightmare to me. Remember too that hoses freeze in the winter and so I would think that overhead hoses would necessarily HAVE to be removed for the winter.

The wash stall in my northern climate barn most definitely should and did have hot and cold water. These lines, as noted earlier, must be run through heated walls in order to keep the water lines from freezing in cold weather. I don't care what anyone tells you, horses shrink from cold water being sprayed on them just the way we do. Cold water is cold—and when you're hot, having cold water sprayed on you is shocking to the entire system and could even be potentially life threatening—horse or human! I can't tell you how many times over my many years of riding and showing that I've apologized to my horses for having to spray them off in cold water or for having to give them a cold water bath.

NOTE: By the way, there is a correct way to spray water on a hot horse. You should start at the back end of the horse, spraying up and down the backside of his back legs, then spraying between his back legs, and up under his tail for as long as it takes to start to see the popped veins in that area of his heated body start receding (several minutes). You next spray the sides of the horse moving slowly up each side and thoroughly wetting each. Next you spray each side of the neck. Lastly, you spray the horse's chest area and between his front legs and under his belly. In this way you avoid spraying cold water onto your hot horse's major organs until the majority of his body is relatively cooled; most especially avoiding his heart area, which if sprayed too soon with cold water, could cause the horse to have a heart attack.

What I can tell you from my personal experience is that my horses now LOVE to have a bath. They just melt into the warm water when I spray it over them. I am grateful every time I bathe them that I no longer have to shock them with cold water, but can now soothe them with warm, cleansing water. Yes, horses DO know the difference, and are very grateful for the warm water. I therefore strongly encourage you to install a freeze proof, hot and cold water "mix" faucet in your wash stall.

Be sure to also consider where your wash rack drain will be draining to! You don't want the wash rack to drain into your septic system in that it would overwhelm the system and potentially clog it up with horse hair. Theoretically, washing off your horse would be equivalent to washing off your car, and therefore the gray water run off from a wash rack should be free to just "run off" across the ground in the same way that car washing water does. However, there might be local restrictions on where such gray water drainage can be drained, and you might want to check that out. Hopefully you will be able to have an underground drain from your wash stall that leads to your drainage pond or other equivalent run-off-friendly area on your property for the gray water to disperse and evaporate—and one where the horses will not be drinking the gray water. If you do have an underground drainage pipe from your wash rack, and the pipes will run under any driveways, then you will want to make sure that the pipe is buried deep enough to be unaffected by any heavy vehicles driving over the driveway so as not to crush the buried pipe. You would also want to make sure that the drainage pipe is below the frost line for your area to avoid freeze ups in the drainage pipe in the winter season.

Photo #40: Wash Stall with Shelves, and Crossties with Sisal Cord

Your wash stall should be lined with bright white, textured plastic, tile board. Bright white is just pleasant and happy. However, it is important to have the waterproof tile board glued onto ¾" plywood backing in order to reduce the sound of the water splashing against the tile board which significantly lowers the spook factor for the horses, and is needed to provide structural support for the tile board. You'll also find it more than handy to have your builder build a set of corner shelves for you in your wash rack. I have my shelves in the back corner next to the water faucet. It is just two shelves that are raised off the floor by about four feet with a hook on the bottom for a bucket and a hook on the first shelf for the hose sprayer. The shelves are made from heavy, treated wood, and can support a lot of weight.

NOTE: Don't forget that your grooming stall and your wash rack will both need crossties. Crossties should be mounted approximately 8 feet, 3 inches above the aisle floor. For 12 foot wide stalls, you should have two cross ties, each one of 5 and ½ feet in length, attached at the top to 20" of breakaway sisal cord. The sisal breakaway cord in turn should be attached to a heavy duty eye bolt screwed into the wall. Even though you have a dedicated grooming stall, it's a good idea to also have one or two sets of cross ties mounted in your barn aisle as extra areas to work with horses.

Having an enclosed **tack room** with a locking man door for security purposes is essential in your barn. I would also highly recommend that the tack room be heated. Not only is a heated room good for the tack, but it allows blankets to fully dry, it keeps shampoos and Show Sheen in usable liquid form in the winter (rather than a useless frozen lump), and it is a "user friendly" place for humans to put on their boots and the like. Recently, I discovered yet another reason why you should have heat in your tack room. There are high quality feeds that I am now using that have a lot of moisture in them that freezes in the cold storage at the mills and are, therefore, sold to you in the winter as a frozen mass. If you don't have a heated room in which to thaw out these frozen grain bags and store the grain in the winter, they will remain frozen in a solid lump and be very difficult to work with. But, because I have the heated tack room just across the aisle from my grain bins, I can conveniently store the grain bags in plastic storage containers (aka, covered, plastic, garbage cans) during the winter in the warmth of the tack room where the grain thaws out and can be easily scooped.

Photo #41: Tack Room with Bridle and Blanket Racks to the Left of Entry Door

Again, my tack room is 12' x 12'—the same size as the stalls. This makes a good sized, basic tack room. There are many configurations you can make within the 12' x 12' area to be more efficient in order to get more "stuff" into the tack room or, conversely, less efficient to give the room more open space. Meanwhile, because 12' x 12' doesn't really mean anything as such on paper, I will tell you how I currently have the room laid out and the amount of tack in it to give you a feel for how much usable space this size room provides.

First of all the walls in the room are fully lined with #2 southern pine, tongue and groove. The ceiling is also fully covered with wood where I used T-111 wood paneling. Using wood on the walls makes a very beautiful tack room as well as a very functional room in that you can securely hang tack holders and blanket holders anywhere on the wall. As you enter through the man door into the tack room, on the walls to the right and the left of the door, I have 10 bridle racks—3 to the right of the door, and 6 to the left. There is room for about 3 more bridle racks if I want them. They are nicely spaced (8″ spacing on centers between the racks) so that the bridles and martingales hang without being squashed. Along the top of the left wall of the room I have four large blanket racks—the kind that fold back against the wall. I have this type of rack so that blankets can be hung fully open and at full length to dry. Theses racks take up a little more than half the length of the wall. Next on that left wall, I have a 75″ high, by 48″ wide, by 18″ deep, moveable wire shelf of commercial strength.

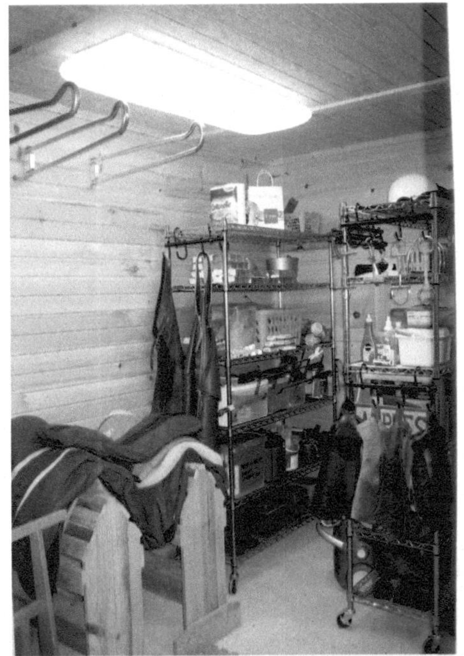

Photo #42: Tack Room Left Wall with Wire Shelves, Blanket Racks, Saddle Stands

NOTE: We put low density spray foam insulation in all the walls behind the tongue and groove wood on the tack room walls and in the ceiling because our tack room is heated. If you plan to heat your tack room, you should insulate the walls and ceiling.

Along the back wall, I have another of the wire shelves that is parallel to the one on the left wall with about three feet of spacing between them. Next to the second wire shelf on the back wall, I have a 72″ high, by 39″ wide, by 21″ deep, storage armoire.

On the right wall, I plan to have three, double rack, wooden, saddle racks mounted on the wall, each rack approximately 22″ L by 21″ H by 7″ W; double mounted height equaling approximately 42″. Next to the saddle racks I plan to hang a couple of wall hooks for extra girths. At the moment my saddles are stored on free standing, saddle stands along the left wall of the tack room that are 24″ wide, by 36″ high, by 9″ wide. On the right wall I have a chest of drawers for storage and a chair for pulling on boots.

Photo #43: Tack Room Back Wall with Armoire and Right Wall with Saddle Stand

As you can see, the walls of the room are full of tack and we haven't even talked about putting anything in the middle of the room. The center floor of the room is open and most of the wall (half way up the wall) along the edges. As I mentioned, I currently have four, free standing saddle racks in the room, plus a couple of large plastic containers with some blankets and Irish knits in them. However, you can be wonderfully creative with the remaining space in the room and do things like put seating in the middle of the room (and the seating could be "made" by using show trunks as benches!) where, for example, you can sit down and put on your boots or change a bit on a bridle.

I think you can see that a 12′ x 12′ tack room is very accommodating. The only thing that would jam up your tack room would be if you wanted to store all your off-season blankets in storage boxes in the tack room or all your show trunks. Then the room would get tight.

In fact, **storage of blankets** is a big issue. You could almost argue that you should have another 12′ x 12′ area just to store the blanket boxes. However, that is a cost consideration. If, however, you have the money to add on an extra "stall" sized area for blanket storage, then you'll be the happier barn owner. But for most of us on a limited budget, lining the aisles with blanket storage boxes works well too.

Also, if you have had your builder reinforce the areas over the enclosed rooms in your barn, then you can put your stored blankets in plastic boxes over those rooms. It's not easy, however, to lift all those boxes up and down from that high storage area, and with the "need" for different weight blankets varying so frequently, it really is much more convenient to have the storage boxes easily accessible from either the aisle or from a separate storage area on ground level. The upper storage areas work better when used for extra, "empty" boxes, and the like, that don't have to be accessed very often.

Let's leave storage areas for a moment.

You will definitely want to have a powder room type **bathroom** in your barn. The key to remember here is to have the bathroom accessible from the aisle. By entering the bath from the aisle, you will cut way down on the dust, dirt and debris that gets tracked into the office/lounge area. You will also get more privacy of use by having a bathroom entered from the aisle.

Photo #44: Barn Bathroom

Inside the bathroom, you will want to be sure to have everything open, so as not to have any "mouse hiding" nooks. In other words, do

NOT have any kind of cabinet under your sink, above your sink, or anywhere in the bathroom. In addition, I have, and recommend you consider having, a pedestal sink that is very attractive, and doesn't require a supporting under cabinet. As far as where extra toilet paper storage goes, buy one of the metal, free standing canisters with a lid and place it on the floor. Just do everything you can to make sure you do not provide any place that allows mice to run and hide. There is nothing more personally disconcerting to me than dropping my breeches in a barn bathroom to then see a mouse go scooting along the baseboard to some mystery hiding zone somewhere near me and my dropped breeches. Yipes.

NOTE: Don't forget to have a ceiling exhaust fan installed in your bathroom!

NOTE: All access man doors to rooms off any barn aisle should be installed to "swing in" to each of the rooms - they should NOT "swing out" into the aisle. I would also recommend that all man doors into rooms be half glass doors (except of course the one for the bathroom door!) so that you can see out into the aisle before exiting the room and make sure no horse or people traffic will be interrupted by your egress.

The bathroom and your office/lounge/viewing area should be your drywall finished rooms. You want them to be bright and happy rooms. Pine is great to use for lining the aisles, stalls, and tack room, but painted drywall and wood trim, like the walls inside your house, are really much more pleasant in the "people" areas, in my opinion. I think having tongue and groove pine as a covering on every barn wall gets too heavy and looks monotonous.

Office/lounge/viewing areas are generally combined into one large area. However, I toured one barn that had a single purpose, viewing area totally separated from an office area and a lounge area rather than a combined office/lounge/viewing room. The separate viewing area was entered through a man door from the aisle on the side opposite from all the other barn activity. Upon entering the room, it had a nice big window, and two levels of stadium like seating for viewing the arena activity. Although on paper, having a dedicated viewing area might have seemed like a great idea to allow for full focus on the horses and riders, the reality of being in that room was quite the opposite. I found that sitting in that isolated, sound-proofed room was the coldest, most isolated feeling imaginable. I almost felt like I had been punished and put in solitary confinement merely because I wanted to watch the horses and riders. I became so consumed by the miserable feel of the room and wanting to leave it, that I had little interest left in watching the horses and riders.

In Amsterdam, the capital of the Netherlands, in an "ancient," but still heavily used and "snobby" stable, they had the complete opposite of the single purpose, observation room. Here they had an open platform raised up right alongside the riding ring for viewing—kind of like a mini deck. There were some chairs placed on the cement platform, with the edge of the platform minimally enclosed by a narrow, wrought iron railing. Although totally open, unlike the isolated, enclosed viewing room of the other barn, this raised platform also made for a very uncomfortable observation area both for our viewing of the horses and riders as they passed by us, as well as for the horses and riders who were "forced" in return to view us sitting right next to them!

Even though I prefer that the office/lounge/viewing room be combined into one area, I feel the combined area needs to be separated into sections. It should not be one combined, totally flat, and open area.

In another barn we toured, the office/lounge/viewing area was HUGE. It was at least a 24'x 24' area. The glass of the viewing window was equivalently huge running from about a foot above ground level, up to at least seven feet in height. That sounds like it should be a totally inspirational room, doesn't it? But it wasn't. Why? Because the whole area was flat and all the "zones" of the room ran uncomfortably into each other. What I mean by that is kids were able to make a full length, chaotic run across and around the room, which encouraged too much expenditure of energy which is totally inappropriate when and where people are riding. In addition, people were cooking and eating all over the room. Then, because the glass was so huge, the tendency was for the sofas and chairs to be quite far back from the glass as if it was assumed you could get a great view of the riders from any point in the room. However, you really couldn't get a good, focused view on the ring because of the kids, and all the food making and eating activities going on in between the sofas and the window. But, on the other hand, if you put chairs up against the window to try and get a better view of the riding, then it was kind of spooky for the horses as well as uncomfortable for the viewer. The open, flat room just made for an overall sloppy, chaotic mess of a room. It had more of a "party, party" feel to it than a room for respectfully observing the efforts of athletes involved in a very highly disciplined activity.

I have also showed at a couple of facilities with huge glass viewing windows that ran the full width of the show ring and had equivalent, erratic and chaotic human activity going on behind the glass. I never felt comfortable showing under those conditions. Most of my horses were surprisingly good about it, all things considered. However, one horse I unfortunately purchased, and subsequently had to sell, would not even consider going past the viewing room glass at those facilities. Over the years I witnessed several negative incidents with other riders and their horses that were clearly caused by human activity behind such wide expanses of glass.

But, now, back to my viewing room and how I planned out the area to hopefully serve as inspiration for your thinking and planning.

Photo #45: Viewing Room and Connector Aisle as Seen from Arena

First, I took a 12′ x 24′ area and divided it into a 12′ x 18′ area for the office/lounge/viewing area and used the remaining 6′ x 12′ for the bathroom and utility room. The actual bathroom size is 5′ x 5′ with the extra area of 2′3″ x 5′ being added to the utility room. By lining up these areas to adjoin one another, it was easy to heat the bathroom along with the office/lounge/viewing area.

Part of the 12′ x 18′ area needed to be used as a utility space. There is a man door in the 12′ x 18′ space that enters into the utility area that takes away about a 3′ 4″ x 9′ 4″ space from the overall 12′ x 18′ foot area. The utility room space is in an L shape which creates a "nook" area on the other side of the wall in the office/lounge/kitchen area.

Photo #46: Covered Entry Porch to Horse Barn Office/Lounge/Viewing Area

There were a few important things that I wanted to have in this area. The first most important thing to me was to have a BIG utility sink with hot and cold water in which I could clean tack—kitty cats--and other things. I also wanted to have a washer and drier in the room so that I could just wash the horse blankets in the barn without having to lug them over to the house. I wanted to have a sitting area where I could also have a desk. Then, of course, I wanted to have a viewing area. I also wanted to be able to enter the room through two separate man doors—one door that entered the room from the interior aisle of the barn, and one door that entered the room from the exterior of the barn from a covered porch area.

NOTE: The office porch is 6 feet deep, 7 and ½ feet wide, and 9 feet high.

And, I wanted **air conditioning in the room in addition to heat**. The heating system in the barn is hot water (antifreeze) in the floors divided into three zones - office & bathroom, tack room and aisles. The heating system is described in more detail in Chapter Eight.

Photo #47: Lounge Area Utility Sink in Nook Area, and Aisle Entry Door

The **utility room** contains the electrical distribution circuit breaker panel for the barn, the softened and unsoftened water distribution valves for water coming from the house to the barn, and a propane fired hot water heater, a heat exchanger to take heat out of hot water and transfer it to the antifreeze in the floors, pumps, and all the plumbing valves and controls for the floor heating system.

Okay, so here's how I planned out the room. The two entrance doors (aisle and exterior) are across from each other and both have the half light doors I suggested. The aisle entry door enters the room "off center" nearer the bathroom side of the overall area. To the right of the door, in the corner, I have the utility tub. This tub DOES have a cabinet above and below it. Utility tubs are just too ugly to have exposed. Besides, I figured I could deal with cabinet mice in this area where my breeches would be full height at all times! Next to the utility tub is where the nook is formed by the utility room. In that area I installed the water connections for a washing machine (covered up in this photo by a piece of unpainted plywood), and the heavy duty electrical service for an electric drier. However, I do not currently have a washer and drier in the room. Instead, at the moment, I have a small cabinet style end table next to the sink cabinet, sided by the kitty litter box, which is then sided by yet another one of those very useful commercial grade, open wire shelves—though this one is only 4 and ½ feet high.

Photo #48: Lounge Area Nook with Washer and Drier Hookups, and Utility Sink

The exterior wall, entry door with the covered porch is directly across the room from the aisle entry door, with the entrance to the utility area to the right of the exterior entry door when looking at that door from the inside of the room. The air conditioner is mounted in a cut out in the wall to the left of the exterior entrance, man door and there is a casement window over the air conditioner. A mini refrigerator is to the right of the air conditioner and window with a microwave oven sitting on top of the fridge. The basic, flat area into which

Photo #49: Lounge Area Exterior Wall Entry Door, and Utility Room Door

Photo #50: Lounge Area Exterior Wall and Exterior Wall Entry Door

you enter is 12′ x 12′ and serves as the basic lounge area.

What I did to keep the areas together, yet separate, is that I took the remaining 6′ x 12′ area that fronts along the arena, which is for the purpose of viewing the horses and riders, and raised the floor of that area. The floor is raised up 2′ 3″ from the flat 12′ x 12′ lounge area. There are three steps up into the raised area to the left of the entry door from the interior aisle. There is a wall that runs across the back of the raised area,

Photo #51: Lounge Area with Viewing Area Cut Out, AC & Exterior Wall Window

separating the viewing area from the office area. The glass of the viewing area (which of course is safety glass—and you have to make sure your builder puts in safety glass because he won't unless you insist and demand that he do as such!) that looks into the arena begins 31 inches above floor level. The viewing glass consists of two glass windows where one, smaller glass window is aligned with the stairs, while the other, larger window is aligned with the actual sit down area of the viewing area. The smaller window is 2′ 8″ wide by 4′ 3″ high. The larger window is 7′ 3″ wide by 4′ 3″ high. The viewing area is just wide enough to fit in chairs for viewing.

Photo #52: Lounge Area to Left of Aisle Entry Door showing Stairs to Viewing Area and Open Wall

At first the builder told my husband that he was making a solid wall between the raised viewing area and the office/lounge/viewing area. I thought to myself that by putting in a solid, blocking wall that it would make the viewing area seem like a closet and make the rest of the office/lounge/viewing area seem small and confined as well. So, I suggested that we open up that separating wall and allow it to be like a "pass through" area you often see in houses where the kitchen backs right onto the dinette. By opening up that wall, leaving the minimal amount of area to support the wall, it made for an open, roomy feel for both areas, yet kept them separated. The opening in the wall is 60 inches wide by 32 inches high.

I should probably note the reason why I don't have a washer and drier in the office area. It's simply a matter of money. I bought the new, large capacity washer and drier and put it in my house. The capacity of my "old" washer and drier was such that the horse blankets didn't fit in freely enough to actually clean them. Because I'm on a budget, I couldn't afford to buy two sets of the new large capacity washers and driers and so, I don't have a set in my barn office area. However, I don't really "miss" having them there. The space where the washer and drier would have gone has been put to good use with the shelving and kitty litter corner. With blanket washing only being an occasional event, I really haven't found it all that inconvenient to drag the blankets over to the house and wash them there. But your budget and desires would determine that call for you!

Photo #53: Lounge Area Aisle Entry Door and Stairs to Viewing Area

By the way, you will probably want to air condition your office/lounge area as I did. Be sure to plan where you want the builder to cut a hole in the wall for the air conditioner. Later I will remind you that the electrician might also have to put in a heavy duty power outlet for the air conditioner if your room will be of a large enough size to require that the air conditioner be large and therefore require a 240 volt outlet.

As I mentioned, I have the **utility room** entrance through a man door from within the office area. Be sure to make the utility room large enough to easily walk into it and be able to work in it. People often fail to plan for a functional utility room which makes it very difficult for service both by you and any workmen. Again, our utility room is an L-shaped room borrowing a little space from the bathroom depth on its back wall and a little space from the shared wall of the office with one long wall of the bathroom. That makes the utility room upon entering, 3'4" wide by 9'4" long, and then ends with a 2'3" x 5' "L" to the right, at the back end. Not only is there plenty of space for functionality within this utility room, but it offers a nice little storage area for extra stuff like XMAS wreaths, vacuum cleaners, and the like.

Photo #54: Horse Barn Entrance Foyer

You need to have **an entrance** into the barn. You can save space and money by entering the barn through the barn aisle end doors or an innocuous man door located near an end door. However, having an actual entrance area—or foyer—in the barn is a lot nicer. In fact, having a "foyer" for your barn is the "fun" part of the barn. It's like the area at a horse show where your barn gets to put up a little "ego and personality corner" (an EPC—and yes, I'm coining and trade-marking that acronym here so please be sure to give me credit for it if you use it!) for passersby to see and admire. Not only can you "dress up" the foyer area with a little style, but having a foyer makes a perfect place to put the grain bins.

I had hoped to have my foyer area highlighted with natural light flowing down into the aisle from an open, glass sided cupola mounted above. Nevertheless, this didn't happen. I settled for my beautiful stall Dutch door that allows a lot of light into the foyer as a front door, and accented the area with a wrought iron chandelier hanging from the foyer ceiling.

But, you might ask why didn't I get a **cupola** on the barn?

Well, first I should clarify that although I didn't get a cupola on the main barn as I had wanted to do, I did put one on the hay barn. I had already ordered the cupola for the hay barn—remember that our hay barn went up first and a long time before the barn and arena construction ever even began—so that I did have the experience of buying and installing cupolas before it would be time to consider putting cupolas on the main barn or arena.

First of all, I was surprised to find that cupolas cost a lot more money than it seems they should cost. But more importantly, the problem with cupolas is that they require cutting a hole in the roof if you want the cupola to allow in light. Once you cut a hole in your roof you invite in a lot of potential problems that, in my opinion, are just not worth it in the long term. Plus, your builder will be very reluctant to cut that hole in the roof, and he will be "challenged" as to how to correctly install a cupola after he's cut the hole. In addition, that little projection atop your barn roof becomes a very efficient lightning attractor such that you will either have to install a lightning grounding system for the cupola or plan on having your barn eventually struck by lightning with a resulting hole in your roof and/or, heaven forbid, a fire! And, yes, I have personally experienced having our ungrounded, garage-top mounted cupola being hit by lightning in our Ohio house and having the lightning blow a hole in the roof!

Photo #55: Hay Barn with Cupola, and Architectural Asphalt Roof Shingles

LESSON #52: Trying to find someone who knows how to ground a lightning rod is outright challenging.

After repeatedly receiving glazed over eyes from discussions with our builder over installing a lightning rod, we finally resorted to looking up how to install a lightning rod online and then telling our carpenter how to install one. We found out the basic system requires that a heavy piece of insulated stranded copper wire be bolted to the copper metal roof at the top of the cupola, and run down the roof, then down the side wall of the barn, and finally into the ground, where the wire is bolted to a buried long copper grounding rod.

Well, I don't know about you, but I had never noticed those big fat wires running down anybody's barn roofs between their cupolas and the ground. But, once I did see the wires, the one thing I knew was that I wasn't about to have any of those ugly black wires draped down across my hay barn roof! Yet, the reality was that I had to have one of those darned wires somewhere if I planned to have a cupola on my roof and if I didn't want to use the hay barn as a landing pad for lightning bolts. So, I had to somehow solve the "ugly black wire" problem which my husband did for me by running the wire from the copper roof of the cupola, down through the ridge vent of the hay barn roof, then down and across the roof trusses, over to the interior side of the exterior barn wall and then down the interior side of the exterior barn wall (but without being near or touching the hay bales in any way), and then out under the exterior wall where the wire was attached to the buried metal rod. Although I can't guarantee that our method is "safe," I can say that we haven't yet gotten struck by lightning—okay, so maybe I should hurry up and knock on wood!

The other huge problem when you cut a hole in the roof, either for a cupola or a skylight, is that you are just daring your roof to leak—and your roof will be more than happy to take you up on your dare. Whereas modern day skylights have some anti-leak installation features (although you will find over time that they are not necessarily, totally leak proof anyway), the traditional cupola does not come with any anti-leak "gear." So you are left with trusting your builder, carpenter or roofer to correctly install, and then flash the cupola. Good luck.

There is another problem with cupolas, as there is with any kind of cut out in a building, whether the hole is on the roof or on the side walls of the structure. Basically it's that to have a hole is to have your choice of any combination of insects, birds, bats or four-legged, fuzzy "critters" using that hole as their entrance and exit door into your barn, and, in the case of a cupola, possibly using it as well for their primary residence. One of the most omnipresent invaders you can expect is bees. You will find that you will have enough trouble "policing" the doors, sides, and edges of your barns for bee and wasp nests. Having to access a cupola to eliminate bee hive problems is trouble that is exponentially more difficult.

To try and ward off the anticipation of bee hive problems in our cupola, we made a point of securely installing screening on the entire interior surface of the cupola when the cupola arrived and was still at ground level. That means that we stapled and glued screening onto the inside of the cupola BEFORE the cupola was installed on the hay barn roof. Trying to put screening in the interior of the cupola AFTER the cupola is installed would be really difficult to do at best—and most probably impossible to do.

At first I was very disappointed about not having a cupola on the main barn. I'm sure you will agree that other than seeing those long runs of beautiful, white, horse fencing lining your pastures, there's nothing that says "horse farm" more than having a cupola on top of the barn. Yet, now that I don't have a cupola on the main barn, and have spent a lot of time looking at cupolas on other people's barns, there is one thing that I have noticed about cupolas in general which may or may not bother you.

What I've noticed is that cupolas have a tendency to date a barn. Cupolas on barns date a barn in the same way that excessive chrome on a car dates a car. Just envision a 1956 Oldsmobile. What says "old" more than all that chrome?

The other common problem people seem to have with cupolas is the same problem that is commonplace with people having trouble putting the correct sized exterior lighting fixtures on their barns and houses. People have a consistent tendency to top their barns with cupolas that are just flat out way too short and small for their barns. I shouldn't fail to mention that there occasionally are those people who will do the opposite and put up cupolas that are way too big and tall for their structure. The result of a cupola being too small or too big can be to make your whole project look tacky and tasteless.

NOTE: The correct way to estimate the height of a cupola is that there should be a 1.5 inch rise in cupola height for every 1 foot of unbroken roofline length of the roof on which the cupola will sit. For example, the 36′ long roof on our hay barn would need a 4 and ½ foot tall cupola (36 x 1.5″ = 54″/12 = 4 and ½ feet). For our horse barn with a 72′ roofline, we would have required a cupola of about 9 feet in height. Remember again that you MUST ground the cupola or it will eventually be hit by lightning and could potentially cause a fire.

Photo #56: Landscape Photo of Horse Facility, and Relative Roof Pitches/Lines

What I did not foresee, but have found over time, is that having a very low profile facility wears well. The low profile flows with the lines of the pasture expanse and makes a natural fit of the entire facility into the surrounding area. That's a really nice thing—especially in the long term.

Now we should return to discussing the **various storage area needs** you will have in your barn other than the storage space you will have above your enclosed rooms.

The first big storage issue is determining where you will put your **shavings**. But before you can consider the answer to that, you must first decide whether you will be getting bulk shavings or bagged shavings.

Once again you will find that you will be at the mercy of a supplier and his delivery service—this time for shavings. The shavings situation is endlessly frustrating. Also, once again, you will find dealing with a supplier opens you up to yet another price playing game. Here is the basic thinking to consider when making your type of shavings choice.

If you have a lot of horses or plan to have a commercial facility, then you would probably find it best to use **bulk shavings**. However, using bulk shavings for your stalls means you will have to build a separate structure in which to store the shavings for safety (fire hazard) and health (dust particulate) reasons at a minimum. But before you build a shavings storage facility, you should first find your bulk shavings supplier and then find out from him what his requirements are for delivering bulk shavings.

Bulk shavings are usually delivered by a truck that has a moving bed floor which means the bottom of the truck has a moving conveyor belt. After having backed the shavings truck into your shavings barn, the shavings are off loaded as the truck moves forward and the conveyor belt pushes the shavings out of the truck. The length and width of your storage building will be determined by your supplier's specific truck. He will also need a specific height for the entrance door into the barn or his truck may not fit into the barn and he won't be able to deliver the shavings.

NOTE: Storing bulk shavings outdoors is not a desirable option. First of all, among many reasons, the shavings can be easily blown around and away. Second, when it rains, the shavings will be wet and backbreakingly heavy to handle when shoveling them into the stalls. Third, wet shavings will freeze into lumps in the winter.

In general, a truck load of bulk shavings is typically 140 cubic yards or 3,780 cubic feet. The building required to accommodate 140 cubic yards of shavings, and the corresponding delivery truck, would be a minimum of 24 feet wide by 48 feet long with side walls at a minimum height of 16 feet. The minimum entry door required on the building would be a door that is 14 feet high by 15 feet wide. The side walls of a steel pole barn structure built to accommodate bulk shavings would need to have the side walls reinforced with 5/8" plywood due to the pressure on the steel walls that the weight of the stored shavings would create.

There are additional problems with the bulk shavings people. One of the most frustrating problems is that there are long periods of time in the spring (especially in northern climates in areas that have four seasons with spring ground frost melting that makes the ground soft) when weight restrictions are imposed on the type of side roads and back roads that generally lead to farms. Weight restrictions mean that your bulk delivery truck will be too heavy to make shavings deliveries during those times and therefore the deliveries won't happen. Don't make the mistake of thinking your supplier will deliver a small load during this time because small loads are not profitable for him. That means you will be left in a shavings crunch if you haven't planned ahead or if you don't have a large enough storage building.

Another problem with bulk shavings is that the quality of the shavings varies tremendously, and at times is substandard. We used to call the worst of the shavings that were delivered to the barn where I boarded "hamster cage quality." The absorbency level of hamster shavings is lower than minimal for horses.

But the biggest problem with bulk shavings is there just aren't many bulk shavings supply people around. Then, of course, as soon as you have limited suppliers, you become quite completely at their mercy when it comes to quality, and pricing, as well as supply. And too, if you should be so unfortunate as to get into any kind of dispute with your supplier, you're basically screwed.

But let's be optimistic and say everything has gone well with your plan to use bulk shavings and you want to use bulk shavings because there appears to be an enormous cost savings over using bagged shavings.

NOTE: The current cost of a truck load of 140 cubic yards of shavings is approximately $2000. One hundred and forty cubic yards of bulk shavings (or 3,780 cubic feet) is roughly equivalent to 687 bags of shavings (5.5 cubic feet of shavings per bag compressed to 3 cubic feet per bag). The current cost of one bag of shavings is approximately $5.00. Therefore the cost of 687 bags of shavings (which is equivalent to one delivery of 140 cubic yards of bulk shavings) is $3500.

Now let's consider the reality of the costs of using bulk shavings. First of all, you will need to construct the shavings storage barn. Remember that stored, bulk shavings also present a fire hazard, just as stored hay does. As such, you will need to construct a separate structure to store the bulk shavings, and the building must be at a minimum separation distance of 50 feet from any live animal structure. In other words, you should not plan to just extend the roof line of any live animal structure such as the horse barn or the riding arena to use as a lean-to storage shed for your bulk shavings—

you need a separate structure. The approximate cost for the minimum sized bulk shavings storage building of 24 feet wide by 48 feet long by 16 feet high is approximately $24,000; not taking into consideration the cost of lining the structure with 5/8" plywood.

Next there is the reality of actually working with bulk shavings. Remember that the shavings are sitting in a storage barn separated from the main barn, and they need to be transported over to the main barn in order to be put into the stalls. How do you plan to do that? And who do you plan to have do that?

The traditional way to transfer the shavings is to have the tractor's front bucket loader scoop up shavings from the storage barn, drive them over to the horse barn, and then dump the shavings in the aisle. Next, the shavings that have been dumped in the aisle have to be shoveled from the aisle into the stall. Not only is this a time consuming event, but it's also a very labor intensive one. So, you must now also add into your savings calculations for bulk shavings a significant labor cost as well as the cost for the gasoline required to run the tractor.

In addition to actual cost figures, all that transferring, dumping, and shoveling of bulk shavings makes for a lot of dust, such that the transfer requires that the end doors of the barn be opened. Having the barn doors open works well in the warm weather, but in the winter, open barn doors make for a cold barn.

Personally, what we found, in addition to all the rest of the considerations of bulk shavings versus bagged, was that, in the end, the difference in cost for a small farm was not sufficiently significant between the two types of shavings to warrant selecting for bulk shavings over bagged shavings especially when compared to the advantages of bagged shavings. Using bulk shavings really only makes sense if you have a large horse facility or a commercial horse facility.

However, relying on **bagged shavings** suppliers isn't nirvana either.

So, if you decide you don't want bulk shavings, and therefore, you won't be building a separate structure to store shavings, you will still have to determine where you will be storing your bagged shavings in your main barn (or other), and how your bagged shavings' delivery truck will access that storage area.

NOTE: Bagged shavings are usually delivered on a trailer pulled by a pick up truck or in the bed of a pick up truck.

First of all you will need to calculate an estimate of how many bags of shavings you will need per month. You will want to make a per month calculation for two reasons: (1) you generally won't be able to conveniently store more than a one month supply; (2)

you won't want to put out more cash outlay than for a one month supply. Shavings are expensive.

You can estimate using a minimum of about one bag of shavings per day per stall in a 12′ x 12′ stall. Your start up amount for that size stall will be between four and six bags per stall. Then after the initial covering, your daily outtake and replacement will average one bag of shavings per day per stall. That means that for every four horses, you will need to have a 12′ x 12′ storage area for 120 bags of shavings. (30 days per month times 4 horses equals 120 bags of shavings).

An open area at the end of a barn aisle next to the barn end doors where the delivery truck can easily pull up and offload the bags is best for bagged shavings. The delivery truck will want to back into your barn aisle alongside the storage area if that's possible and you have "allowed" for it. You will quickly learn that it is to your advantage to keep all your delivery and service people "happy." In this case, that means you should be sure to plan a way for the shavings people to be able to back into the barn. However, letting them back into the barn always makes me very nervous, especially in winter when the driveway right outside the barn door is icy.

Getting a supply of bagged shavings isn't always easy either. First of all, there is an equivalent delivery problem for the mills getting bagged shavings in the late winter/early spring. I'm not sure if the "problem" arises from the same roadway weight restrictions at that time of year or not. It would seem to me that logically with most mills being located on main roads that they wouldn't fall victim to weight restrictions on delivery trucks. Considering the poor quality of the bulk shavings I have seen at that time of year—the hamster shavings—it seems to me that the problem might be more of a source supply problem at the saw mill. It may just be that the saw mills don't have any quality shavings or any shavings at all to bag at that time of year.

Photo #57: Bagged Shavings Storage Area

NOTE: The consistent quality of bagged shavings is more reliable than bulk shavings. I've had very little variation in quality with my bagged shavings.

But whatever the source of the problem, there is a consistent problem of getting bagged shavings in late February and late March (again, in places with four seasons). Be warned and maybe plan to "afford" to buy and be able to store at least an extra month's worth of shavings for that time period. This is an occasion when that empty stall can come in handy.

The other consistent problem of getting bagged shavings is the strangely aloof attitude of mill owners. Why that attitude exists I just don't know. Nevertheless mill owners consistently act as if a regular customer is an inconvenience to them. Go figure. I'm often left thinking that it must be some kind of a power trip for them to make it difficult for you to get shavings knowing that without them your farm is basically dysfunctional. They have a complete circle of behavior that gets you coming and going. They make it difficult for you to schedule a delivery, they are "iffy" about whether or not they will be able to supply the amount of shavings you need, and they are judgmental about how many bags of shavings you are using. Can you explain to me why they would care if I use more shavings than someone else? Why wouldn't it just make them happy that they are consistently making more money off me?

Then there's another "game" that mills play which is to refuse to keep an inventory of bagged shavings. What they try to do is play an "order when they get an order game" to minimize their inventory. But then in response, the supplier to the mill reacts to the mill's constant "small" orders by not making shipments until they have enough "small" orders to make it worthwhile for them to make one big delivery. Meanwhile, it's the farm owner who is the loser in this game. I actually had one of the mills ask me when I complained that I was out of shavings and the mill didn't have any to sell me—with a very nasty, authoritative tone—why I hadn't planned out better for when I needed shavings? Excuse me? Shouldn't the question be why they hadn't planned out better for when their customers would need shavings and then kept an inventory on hand so they could supply us? If money was no object, I could see justifying building a separate structure just to storehouse a large amount of bagged shavings in order to avoid these situations as much as possible.

Anyway, the point is to shop around and try to find the friendliest mill to deal with, and then keep at least a two week extra supply on hand, order two weeks before you really need shavings, and get a regular supply from more than one mill—in other words, play the mills off against each other.

NOTE: Clearly you can see the importance of being near horse goods suppliers such as mills for shavings and feed. That brings to mind the importance of building your horse farm in a horse friendly community, i.e. a community where there are a lot of other horse farms. If you are in an area without horse farms, getting supplies for your horses will be difficult, if not close to impossible, and expensive.

The day that finally sent me out looking for another mill to be my primary supplier was when I ordered shavings on Monday to be delivered on Friday. When Friday came and went and no shavings were delivered, I called to find out that the mill had sold my shavings out from under me to someone else and there were no more shavings for sale at the mill, nor would there be any for at least a week, if not longer. I was really angry.

Nevertheless, even with all the games and trouble, I still do love the convenience of bagged shavings and would be unhappy the day I might have to deal with bulk shavings.

NOTE: Bagged shavings come as 45 bags on a pallet. Mill owners like it if you order your bagged shavings by the pallet because a pallet already has the 45 bags shrink wrapped together and are on a wood pallet from their delivery to the mill. The prepackaged pallet of shavings is then easily lifted by the mill's forklift onto their truck bed or trailer bed for delivery to you. So when you do your bagged shavings order, try and calculate your need and order by the number of pallets you need rather than the number of bags.

There is one last note I should mention regarding bagged shavings. It is possible to order a full tractor trailer load of bagged shavings directly from the manufacturer to be delivered to your farm. By doing this, you will save a lot of money and have a huge supply on hand. However, there are two catches to this type of delivery. First is the need for some place to store all those bagged shavings. However, if you are buying the type of bagged shavings that come in individually packed, plastic bags, then you can theoretically store the bags somewhere outside—although that would be rather unattractive and an invitation for rodents to use as a mouse hotel.

NOTE: Bagged shavings that come bagged in paper will freeze into solid lumps in the winter in northern climates if exposed to any moisture and therefore cannot be practically stored outside.

The second catch to a tractor trailer delivery to your farm is that it is YOUR responsibility to offload all the bags of shavings. Further, you must offload all of the bags of shavings within a two hour time period as the truck driver waits for you to finish. This will require a forklift truck or a heavy duty tractor with a fork lift capability that can carry a heavy weight load to be able to do this. Generally speaking, your average compact utility tractor with a fork lift attachment will **not** be able to off load the weight of a pallet of bagged shavings and may not be able to lift high enough to do the job either. You must check with your tractor dealer to find out the lift and weight limitations of your proposed tractor and fork lift attachment or a fork lift truck in relation to the requirements for offloading the particular tractor trailer you will have delivering a full load of bagged shavings to your farm.

Lastly, you should know that you don't have to have the mill deliver your bagged shavings. You can drive your own pick up truck or trailer to the mill and have the mill workers load the number of bags you want, and then stack them yourself in your shavings storage area on your farm.

There are still three other **storage areas** you will need.

One small area is needed for **wheelbarrows, muck rakes, brooms, and dust pans**. There are a couple of ways to handle storage of this smaller equipment. You can build a narrow (3' x 12') open-ended nook or you can just hang or lean your rakes et al along an aisle wall and park your wheelbarrows in the aisles. The rakes and brooms and wheelbarrow are in such constant use that they will be in the aisle most of the day anyway, so having a dedicated storage area for them isn't really necessary.

The other storage area needed is for your **"short term" hay supply**. Even though you have a nice, big hay barn with a year's worth of hay, you won't want to be going over to that barn on a regular basis to get your daily hay. You will eventually come to find it burdensome to go to the "big" hay barn even every four days! Anyway, the point is that you will need a place in your barn to store a few days supply of hay.

Probably what first came to your mind when I mentioned the need for a few days supply of hay bales within the barn was the visual of a hay cart sitting in the barn aisle piled high with hay bales. In fact, that's yet another one of those visual icons of horse barns. It is also one with which I tried to work on a regular, functional basis, but have come to learn is not the way to store your short term hay. Let me explain.

Our new house here on the farm is significantly smaller than our previous house. Even though I had worked to eliminate a lot of the "stuff" I had in the bigger house before moving into the new one, I missed the mark by about 100 boxes! That meant that for the first several months, the arena served as a storage center for the "extra" boxes until I could find the time and energy to sort through all 100 of them.

In the process of sorting, I stored some of the items in the empty 4th stall which were eventually meant to be put in the storage area above the enclosed rooms in the barn. Although I managed to clean out the arena boxes by mid October, the boxes labeled for storage that were temporarily put in the 4th stall, had still not been put into the upper storage area by December. With so many other "fun" diversions going on at the farm, like the collapsing stall floors, and the frozen automatic waterers, neither my husband nor I were exactly focused on getting those storage boxes moved from the 4th stall to the upper storage area.

It was around Christmas time when I was mucking out stalls that I first caught sight of something out of the corner of my eye scooting along the barn aisle, and then slipping out under the end door of the barn. Unfortunately my brain told me that what my eye had seen was small, dark gray, and fuzzy. As much as my mind wanted to tell my eye that it hadn't processed the visual correctly, I knew, deep in my sense of reality that mice had moved in.

I had also noted that my horse in the 3rd stall next to the 4th stall "storage area" was acting all "jumpy" which was completely uncharacteristic of his normally mellow soul. Nevertheless, I "chose" for the moment to decide that the "spotting" of the mouse was a random event even though I was fully aware of the saying that "for every one mouse you see, there are about a hundred more you don't see." Yet, denial is always such an easy and pleasant path to follow. Unfortunately, denial is also ultimately the path of fools.

Over the next couple of weeks, the scooting fuzzy gray ball sightings increased. Finally I decided to make a quick, exploratory adventure into the 4th stall just to get a visual of the situation. Unfortunately there were definite signs of the 4th stall having become, not just a place of mouse hotel accommodations, but a mouse resort. Plus my horse in the third stall had become jumpier than ever. There was no question now that we had a "mouse problem" and something had to be done about it. We clearly needed to get the next classic icon of a farm—the **barn cat**!

When I told my husband about the mouse problem and that we needed to get a barn cat, he immediately balked and said "no cat!" He informed me that he would take care of the mouse situation by simply putting down rat poison. Yipes! Can you think of anything more inappropriate to do in a barn full of animals and feed? I can't!

NOTE: Whereas rat poison is bad in the horse barn and should never be used in there, you might need to consider putting down **rat poison bars** in the hay barn. You may find that you have a hay barn rodent problem and unfortunately, putting down poison could turn out to be your only choice to solve the problem. However, the hay barn poison should only be in the bar form (not loose grains or powders—even if boxed in containers because it is easily spread and dropped) and placed along the back walls of the hay barn on floor level, under the edge of the pallets, behind bales of hay so that the bales of hay are blocking access to the poison by pets, as are the wood pallets. If you choose to put down rat poison in the hay barn, then you **must always be obsessively vigilant to never allow your pets in the hay barn or leave the hay barn doors open!** You would need to plan ahead and place the poison bars right before your hay is delivered—right after you've cleaned out all droppings of hay, etc., from last year. You would **never** put down poison without stacks of hay bales to block accidental, brief access into the barn by your pets. The rat poison bars should keep rodent populations under control for a year at a time.

Meanwhile, regarding a cat, we argued. In fact, we argued a lot! What you need to understand is that my husband is fiercely allergic to cats and he was fighting for what he felt was his right of entrance into the barn and a commensurate ability to breathe once he was there. Okay, he had a point. His objection to cats was understandable. However, on the other hand, there was absolutely no way we were going to put down rat poison in the horse barn!

To my mind, the reality remained constant in a simple "1, 2, 3, go" kind of way. A barn cat was the only solution and that was that!

(1) We had a mouse problem.
(2) We weren't putting down rat poison.
(3) We were going to get cats, and the cats would get the mice!

So it was clear to me that the only logical thing to do in order to solve the rodent problem was for me to completely ignore my husband. I think any of you women in riding reading this will fully understand that this technique is one we equestrian women easily and readily use on a consistent basis. Sure, try and tell me you haven't ever hidden the "real costs" of riding from your husbands. So, with the "standard operating procedure" of a female equestrian, I simply proceeded to contact a friend of mine who had asked me if I wanted some barn kittens way back in July of the first year we moved in.

Luckily my friend still had the kittens—who were now cats—and still wanted to find three of them a home. I had only planned on getting one cat and was quite certain I didn't want three. What I did get were two adorable cats—a brother and sister—one calico and one tiger cat. So, I put the cats in the barn, and ...

Enter the cats—exit the mice!

Yes, cats do their jobs and they do it quickly. We haven't had any problems with mice since the cats arrived—unless you consider scooping up the dead mice from the aisle to be a mouse "problem." But even finding dead mice is rare at this point in time. The other good news is that my husband can fully breathe in the barn! In fact he loves the cats and even considers them to be his! Apparently his cat allergies have subsided.

NOTE: Don't forget about **farm dogs** too! For the same reason your neighbors have "big dogs" you should also have big farm dogs. Dogs are not only your friends, but they are one of the best advance warning systems available. They also provide a great sense of security when walking around your property—both day and night—against any type of invader, be it wild life or human. Whereas having one big farm dog is good, having two big farm dogs is significantly better, in my opinion.

Anyway, I digress.

So now we had the cats and no mice, but a new problem cropped up. Shortly after the arrival of the cats, the horses started refusing to eat their hay. They would pick at it and eat some of it, but overall the horses were rejecting the hay.

My first thought was mold. I thoroughly checked the bales for mold, but, no, it wasn't mold. I then considered all kinds of other things that could be the source of the problem, but upon investigating, none of them panned out.

As I pondered the many potential sources of the new "hay problem," the cats sat contently watching me from atop the hay bales on the hay wagon that was parked in the aisle for short term hay storage as I followed along my dead end paths of reasoning. Then finally one day when I looked back at the cats, who were looking back at me from their hay wagon vantage point, my mind went ….

BINGO!

I suddenly realized there was a high probability that our little feline darlings were using the short term hay bales on the wagon as an annex to their kitty litter box! No wonder the horses weren't eating the hay. I wouldn't eat it either!

Armed now with this new insight, we realized that we couldn't leave hay out in the open—not even for the short term. We would have to find some way to contain the short term hay that would make it inaccessible to the cats.

Photo #58: Short Term, In-Barn, Hay Storage Building

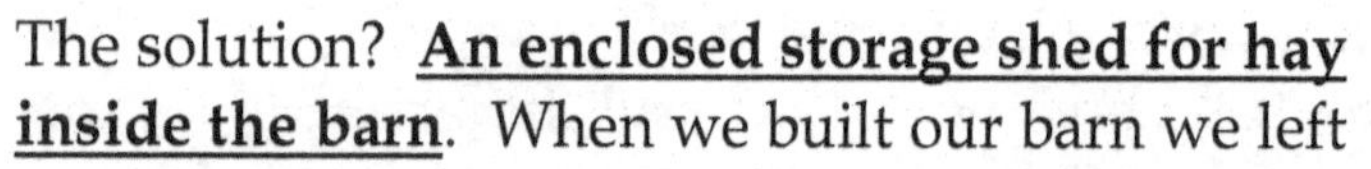

The solution? **An enclosed storage shed for hay inside the barn**. When we built our barn we left a long open unfinished area where "future" stalls could be built along one side of the barn aisle. The area was already serving as our shavings storage area, and light weight equipment storage area. Because the area is so long (12' x 36') we also had plenty of room to put in one of those plastic storage buildings people use for outside equipment storage in their yards. We bought one that looks like a little barn and has nice big entrance doors and put the "little" hay barn right in that open, unfinished area of the horse barn for our short term hay storage. This has worked out perfectly. However, come the day we need to finish off the extra stalls, we will have to rethink things!

In planning your barn, you could do as we did and put a plastic storage building somewhere within your barn. But having a heads up for the need, you could also plan ahead and make a dedicated, enclosed area for your short term hay storage by building either a hay "closet" or "bin"--either of which should be lined with metal to minimize any potential fire hazard. Having an enclosed area for your short term hay also keeps the hay from getting dusty, as well as from mice getting into, and between, the bales before they are caught—meow!

NOTE: As I previously mentioned, **dust** is an ongoing battle in barns. No matter how hard you try to minimize it or eliminate it, dust happens—and lots of it—every second of every day! You can dust off the tops of your show trunks stored in the aisle one day, and be assured that they will be dust covered within 24 hours, disgustingly so in 3 days, and embarrassingly so in a week. You will eventually learn "not to see" dust because you just won't have the time or energy to keep up with dusting, especially if you are operating the farm by yourself. On the other hand, you don't want to be feeding any of that dust to your horses. Therefore, the need for a temporary plastic hay storage barn or an equivalent enclosed area for your short term hay storage in your barn is necessary to keep the hay dust free.

By the way, the cats were the problem with the horses not eating the hay which was confirmed once the temporary hay storage barn was in use and the hay wagon was retired!

Of course if you have your hay storage above your barn (attic area), then you will be doing a "hay drop" down into the aisle and the preceding section won't be important to you. However, once again, I hope that you will be able to have a separate structure for your hay both for reasons of fire safety, and for minimizing the dust and pollutants dropping down onto the horses from the above storage hay area and compromising having good, clean air in the barn for the horses to breathe.

Last you will need an area for **grain storage**. I've been in barns where an enclosed room entered by a man door was used for grain storage, and the grain bags where just laid inside on the floor. This is not a good idea. Mice WILL get into that room and eat into the bags and contaminate the feed. Worse, the barn where I boarded had open grain bins with the feed sitting in them. I would just cringe when I walked into the grain room and would see those open bins. I had to fight to keep my mind from envisioning just how many mice must be contaminating the feed at night in that open feed bin. Ugh. That boarding facility also had a grain hopper into which a grain truck delivered bulk grain which was then dispensed into an open, grain push cart and distributed to the horses. But, of course, too much grain was always dispensed into the cart, and the remainder from the feedings was just left sitting over night in the open cart. Ugh, again!

That brings to mind the point that if you are going to build a commercial facility, and need to consider bulk grain delivery, and you plan to try and have "one grain that fits all," then you will want to install a grain hopper on the outside of your barn for bulk grain delivery. You will also have to make sure that your grain "room" is located in a place that can allow a truck to easily access the outside grain hopper for delivery.

Photo #59: Grain Storage Bins in Horse Barn Foyer

However, if you are going to have bagged feed and additives—which allow you to have variety to meet the different feed and supplement needs of the different horses—then you will need to have **enclosed** grain bins.

Some people just use metal trash cans in which to store their feed bags to make them mouse proof and that generally works okay. However, because you are constructing a barn, you should plan a "grain bin area" into your barn to make it easier for you to feed the horses, and to make a more space efficient place to store grain. As I mentioned earlier, the foyer is a nice, convenient, and pleasant place to have your grain bins. The grain bins can be made from the same pine, tongue and groove wood used to cover the barn walls, with top opening lids. If you can convince someone to line the grain bins with tin and smooth, overlapping edges, you'll be even better off. However, a nice, tight fitting of the wood with metal trim on the edges to match your barn, and "tucking in" of the corners and edges of the bins to make them smooth and air tight, should keep the bins mouse proof. You can also put covered containers inside the bins if there is a mouse problem anyway—but that is unlikely to happen between the bins being located in the openness of the foyer, the basic tightness of the bins, and the presence of the cats. You'll want to have the lid of your bins either have hydraulic closers on them or have a hook and cable system so that you can open

Photo #60: Grain Storage Bin with Open Lid, Hydraulic Lift, and Cable Attachment

the heavy lids and secure them while you are working in them to keep them from slamming down closed onto your fingers. The tongue and groove covered lids are VERY heavy!

__NOTE:__ My grain bins are 2′ deep by 8′ long, divided into 2 bins, each 2′ deep by 4′ wide. The front edge height is 2′8″. The bins slope up to a height of 3′4″ along the back edge.

Just a note again that I now also have the "annex" to the aisle feed bins (tall, plastic trash cans) that is located in the heated tack room because of the freezing nature (in the winter) of the feed I am now using.

__NOTE:__ Your grain bags have information on them that will tell you how much feed to give your horses. You will need a scale with which to weigh your feed. Remember that you should bring feed from your horses' previous facility and gradually transition the horses to your new grain (if you will be feeding grain different from the previous barn) so as not to upset the horses' digestive system. Once your horses are on their new grain, monitor their energy level to determine if you need to adjust their grain either up or down for the horses to have the appropriate energy level.

__NOTE:__ Be sure to remember that horses need to be __regularly wormed__. You can either choose to worm every other month with an oral paste containing ivermectin (least effective method and not generally recommended because not all types of worms are killed by ivermectin) or do a rotation cycle (which you can find online) of various oral paste worming medications that contain different ingredients that attack and kill several different types of worms. Or you can choose to use a daily wormer for your horses which gives the horses a constant low dose of worming medication. If you choose to use a daily wormer, then you must give a boost to the daily wormer by giving one dose of oral paste ivermectin in the spring of the year, and one dose of oral paste ivermectin in the fall of the year. Lastly, you can choose to have your veterinarian do the worming for your horses. You should know that horse people all have strong, individual opinions as to which method they believe is correct for worming.

__NOTE:__ Another purported benefit of using daily wormer is that it is considered to be a colic preventative. As I understand it, the manufacturer of the daily worming product will guarantee your horse will not colic if you start using their daily wormer product before the horse is of a certain age (about ten years of age) and continue to use it every day. If your horse does colic, they will supposedly pay for the colic surgery. I'm sure there are many restrictions and qualifiers which you need to find out about from the manufacturer to take advantage of this program. What I have noticed in my personal experience of using the daily wormer for our horses on our farm is that my horses now have no inclination toward even a pre-colic state as they occasionally did at the old boarding facility where they were not on daily worming medicine. The only pre-colic incident I've had on our farm was thanks to Mr. Bobblehead and the worm-filled apples!

Phew! That was a LOT of information! It was also a lot of floor space!

Let me review how we handled all these storage areas to keep the cost of all this extra space down. When we planned our barn, we allowed for our immediate need of stalls for our horses and fully finished off those stall areas. Then we planned for "future" stalls across the aisle from the finished stalls by putting in the stall footing, the stall windows, and the stall lights, as well as finishing the back of the walls of the future stalls in tongue and groove pine. However, we did **not** put up the stall separating walls or the stall fronts. In this way, as I mentioned earlier, we have a 12′ x 36′ foot open area in the barn that serves as the storage area for the bagged shavings area at one end, the short term hay storage in the middle, and the light equipment storage at the other end.

Photo #61: Open Storage Area

Planning for these storage areas is essential. The first time our hay farmer came into our barn he commented as follows: "Oh, my God! Finally someone who planned for storage in the barn! I can't tell you how many people forget to plan for storage and are screwed!" That kind of says it all. I should probably also mention that our hay farmer was equivalently impressed by our waterers saying, "Finally, a horse person who has installed automatic waterers!" I have to agree that he was right in noting the rarity of automatic waterers in horse barns, in that none of the show barns where we showed, and none of the barns where we toured before building (save one), nor the barn where we boarded, had automatic waterers.

The hay farmer then digressed into talking about how in the summer his pigs actually wedge themselves under the automatic waterers in his barn to make a constant flow of water under which they cool themselves off. What a visual to imagine. And you thought pigs were dumb!

Photo #62: Horse Barn Front Porch

So far we have planned and envisioned the barn in two dimensions by drawing the floor plan on paper and doing linear sizing of the various areas. Let's now take into consideration the **height of the exterior walls** for your barn and get a more three-dimensional feel for your structures.

NOTE: Now that you are considering the barn in three dimensions, consider putting a porch on your barn. Having a porch across the front of your horse stalls allows the horse to have his stall window open for ventilation, while being shaded from the sun and protected from the rain. Porches on barns are also very attractive and greatly enhance the overall beauty of the facility. Remember too that if you plan to have an exterior entrance door into your office/lounge/viewing area that you should also have a covered porch there as well to protect you from the weather when trying to unlock the door, if nothing else. **(see Photo #46)**

Photo #63: Horse Barn Front Porch Side View

The height of the exterior walls is a crucially important consideration at a minimum, for: (a) truck deliveries into the barn or arena; (b) tractor and arena drag access to both the barn and arena; (c) jumping horses in the arena. But of equivalent importance, in my opinion, is the "pleasantness factor" of having high walls in both the barn and the arena. As far as I'm concerned, raising the height of the exterior walls is the least expensive thing you can do to maximize your happiness and that of your horses on a daily basis.

High ceilings are not only visually pleasant, but they allow in more light to the barn, as well as better air flow through the barn.

Photo #64: Main Barn Aisle, Suspended Lights, Eave Ceiling, Glass End Doors

I'm pretty sure that old barns were built with low ceilings to keep in the heat in the winter season so the animals would stay warm. I'd imagine the old thinking of having hay storage above the barn aisle was meant to afford the extra advantage of trapping additional heat for the animals in winter below the hay-bale insulated, aisle ceiling. However, we've already talked about the current realization that the dust and droppings from "above aisle" hay storage is actually a negative for your horses. You've probably noticed that many old barns were also built into the ground and were most probably done that way so as to allow for a kind of passive, natural geothermal, year round, ground temperature of 55 degrees to be maintained in the barn, thereby keeping the barns cool in the summer and relatively warm in the winter. However, in this day and age—for horses—where blankets are readily available in a full range of warmth and pricing, it seems to me that appropriately blanketing your horse in the winter to keep him warm under high ceilings is better than the "old" alternative. But, again, it is your barn, and, of course, your choice.

The exterior walls of our horse barn are 11 feet high. Combining the roof pitch of the barn which is a 6:12 pitch (meaning that for every 12 feet in width, there is a 6 foot rise in height), and the exterior wall heights, that makes the total height of the aisle ceiling in the middle of the aisle 20 feet above the barn aisle floor. This makes for a very pretty, very airy, and very pleasant barn interior.

NOTE: The importance of air flow cannot be over emphasized. In fact, barns MUST have a lot of air flow in that horse waste products create pollutants in the air which are bad for the horses' lungs and health. This is why having a barn that is too "tight" is bad. It is actually better for a barn to be cold in the winter from the outside air flowing in, than it is to have a barn that is closed up to be warm, with consequent stale air.

Pole barn exterior walls have venting along the top edge of the wall. You will note in pole barns you tour before your construction begins that there is a gap between the top of the exterior wall, sheet metal edge, and the roof edge that is left open to air and covered with an open, mesh material. This is your barn's constant, minimal supply of air flow.

It is important to note that if you put a porch on your barn, the venting along the top edge of the exterior wall, where the porch roof joins the exterior wall of the barn, gets eliminated. Therefore, in those areas where the roof porch cuts off this venting, you must insist upon having air vents (rectangular cut outs about 6" x 1' covered with slatted, metal, venting covers) cut into the wall above each stall's back wall about every 12 feet (centered between the 12 foot spacing of the support poles) to access the porch roof's "attic" so as to allow air flow between the barn and the "attic" of the porch roof. Otherwise in summer time, the temperature builds up in the porch roof attic and consequently heats up the barn wall, which in turn makes the barn temperature rise like an oven. There must be air flow between the barn and the porch roof attic to allow temperature moderation (temperature equilibrium) to occur.

Photo #65: Porch Attic Interior Wall Vents, Stall Separator Walls, Stall Ceiling Lights

Even with the mesh covered air vents along the top of the exterior walls of your barn, air flow needs a boost in hot climates or in temperate zones during the heat of the summer. Our little "southern belle" barn designer had automatically included **ceiling fans** in her barn rendition for us. She had nicely placed one ceiling fan above each stall in her drawings. My first reaction when I saw the neat row of ceiling fans drawn on the barn plan was that the horses would hit their heads on them. But then it was pointed out to me that because the barn ceilings were high, the fans would subsequently be placed sufficiently high above the stalls so that head hitting would not be an issue.

I wasn't fully convinced that the fan information was true, and I also worried about other things happening with the fans like having the fans fall off the ceilings onto the horses. Nevertheless with "Miss South" and my husband feeling so sure that ceiling fans weren't dangerous for the horses, I felt obliged to at least consider having them. So, I proceeded to envision how nice it would be for the horses to have the fans in the summer heat to cool them, as well as to potentially have the fans create enough breeze to disperse any fly traffic in the stalls.

Yet, somehow along the way of barn planning, the whole ceiling fan concept was completely lost. I have to tell you though, now that I am in the barn and have witnessed first hand the phenomenal amount of dust that continually coats EVERYTHING in the barn, that I am very happy that we never did install any ceiling fans.

NOTE: You have noticed how many times I have mentioned dust. There is one other important note to make about dust. Be sure that you always have a **full, floor to ceiling wall separating any horse stall areas from any arena/riding areas**. The amount of dust that is created by horses being ridden on arena sand is too much for the health of your horses' lungs. There are some barns I have seen where the front of the horse stalls open directly onto the arena or onto a path (with or without a low wall) around the arena. This is BAD for your horses' health.

Thankfully, however, I had made sure that I had an electrical circuit and outlet box on each stall in order to "hang" fans directly on the stall fronts. Hanging fans on the stalls has worked out just great. In fact, because I have open grills on the swing door, I hung box fans both "up" on the "top" of the right side of the stall front as well as "low" on the door grill—in other words, two box fans per stall the first summer. In the summer of our first year in the barn, when the heat was awful, we also put HUGE, 42" belt drive drum fans at either end of the barn aisle which formed a wind tunnel effect through the barn. Eventually, we bought more drum fans and eliminated the lower box fans on the stall doors. I should mention that, by the end of the summer, all the stall box fans are garbage can worthy due to the impenetrable coating of dust and gunk that has combined and solidified on them. Can you imagine what permanently affixed ceiling fans would look like?

Remember the $14 million house and horse farm we toured? The one where the architect/builder, "I want $10,000 to 'look' at your property" leech presented himself to us? Well, amazingly we were subsequently brought back to that very same barn by one of our potential barn builder bidders (the one who quoted us the most outrageous pricing) in that he claimed to have built that barn. When we went into the millionaire's barn with our prospective barn builder one cold, winter day to see this builder's work, we were shocked to find that the interior of the millionaire's barn was at least as cold, if not even colder, than outside the barn. There was also a wind tunnel effect occurring through the barn (remember this was winter--not summer!) to the point that we questioned the builder about the wind tunnel in that it was clearly making an already cold barn exponentially colder. The prospective builder told us in response that the owner had actually gone out of his way to specify an excessive amount of air flow by opening up the soffits on each end of the barn over the barn's end doors as well as having the traditional opening of the soffits along the top edge of the exterior walls. This additional opening over the end doors had created the wind tunnel effect which the owner insisted was necessary for the health of his horses.

Well maybe that is a point that could be argued, but for us it seemed more like total "air flow overdo," as we stood there shivering in the barn.

The only reason that would seem to justify the allowance by this man for such an extreme air flow through his barn was the fact that his horses were all in Florida every winter for the winter polo season. It would make sense that if your horses were only in the barn in the summer, that you would want to try and make the biggest amount of air flow possible through your barn. You wouldn't care how cold the air flow would make the barn in the winter because your horses weren't there. Well, that is, none of his horses were there for the winter except for the one, poor old horse left behind and blanketless whom we saw standing miserably cold outside, having used every bit of his energy to grow a coat the thickness of which I will never forget. Let's just say that the coat of a wooly mammoth would have nothing on this horse.

I should also offer as a side note—and as support for my general opinion of barn builders—why I mentioned that this builder had "claimed" to have built the millionaire's barn. While we were touring the barn with the prospective builder, the owner of the farm suddenly came zipping down to the barn on his Gator, wildly pushing through the several inches of fresh snowfall, shot gun in tow, clearly ready to fend off the invaders who had infiltrated his property (noting that the owner was apparently not in Florida at the moment). Seeing the owner's wild approach, the builder excused himself, and left us in the barn aisle so he could intercept the owner at the barn entrance door. What followed was a tensely uncomfortable and embarrassing exchange between the "builder" and the owner which we could over hear and during which the farm owner was trying to "remember" who this guy was who was "claiming" to have been his barn builder. Our prospective barn builder kept saying over and over again how he had worked for Bailey Builders and how he had cleared bringing us (his customers) to the barn today with the millionaire's farm foreman, Ron.

"Oh, yes," the owner finally said with sudden, bemused recognition, "I remember you now. You used to be a lot fatter then," he chuckled. "Yeah, you worked for Joe. I remember you now," he chuckled again.

It had now been revealed to us by chance that this "builder" clearly hadn't been THE builder for this horse farm at all, but was merely one of the workers on the project. Not that there's anything wrong with someone trying to start their own barn building business by bending the truth a little here and there to look a little bigger and better than they currently are in order to get your business. Nevertheless, this example offers you another "heads up" on what you will be running into while trying to find a barn builder. Remember that in the end, this "builder" gave us one of the most outrageous quotes.

Returning to the subject at hand, exterior wall heights, I have 11 foot exterior walls in the barn, as I mentioned. Upon entering the barn, everyone invariably breathes in and blinks as if they suddenly feel "inspired" by something. The gasp of enlightenment is then invariably followed by a comment on how much they LOVE the "feel" of the barn. Each time I watch the reaction, I can tell that the people can't quite put their finger on why it feels so good to stand in our barn, but I know that it's because of the high walls and the subsequent high ceilings with the center height of the aisle roof being at an unobstructed 20 feet.

NOTE: I paid extra in the barn to have a **rafter ceiling versus trusses**. Rafters make a clear, open expanse from the floor to the roof of the barn like an open, vaulted ceiling in a house.

The 6:12 roof pitch also gives a very pretty interior line as well as exterior line to the barn. If you want a more dramatic roof line, you can increase your roof pitch. Alternatively, if you want to save a little money in roofing materials, you can lower the roof pitch, however, the attractiveness of your buildings will suffer greatly as well as the interior "feel" of your barn.

Again, having a pitch to your roof, as versus a flat roof, is important to allow the snow to slide off in northern climates, and to allow the water from melting snow to run off. Even if you live in a year-round warm climate, you should have a pitched roof to allow rain water to run off. A flat roof, conversely, allows water to pond on the roof which will eventually cause your roof to leak.

NOTE: Don't forget that all the run off from the roof should be flowing down into your gutters and then down into your downspouts that have long, above ground concentrator/diffuser pipes at the end of each downspout to disperse the water as far away from your buildings as possible!

Now that you have determined your exterior wall height and roof pitch, your "flat" barn can now be envisioned three dimensionally.

As long as the subject of roofs has come up, we might as well finish the thought with a discussion about **roofing materials** in that the type of roof you want will be an early decision for you to make.

Basically there are two types of roofing materials for you to consider.

Assuming you are considering building a steel pole barn, your pole barn supplier will want to put a steel roof on the barn. In fact, that will probably be the type of roofing material that is in your basic quote. However, the supplier may say that the quote gives you the option of having a roof "**ready**" for asphalt shingles" for the same price. Be careful here.

When I hear the words **metal roof**, several things come to my mind.

The first thing I think of is a severe thunderstorm in Florida while visiting friends who lived in a metal roofed house. I was shocked at how the sound of the rain on their metal roof was just deafening during that storm. Remember that horses have bigger ears than we do and their ears are more sensitive to sound than our ears. In addition--unlike us—horses unfortunately do not have the ability to cover their ears and shut out or muffle the horrific sound of such deafening rain. So as bad as the noise of a metal roof in a rainstorm may be for us humans to tolerate, it is exponentially worse for a horse.

The second thing I think of with a metal roof is rust. I think of all the old metal buildings I have passed through the years and noted how over time, the metal roofs have rusted and become flat out ugly.

The third thing that I think of is water dripping down from the ceiling. I haven't ridden under a metal roof yet that hasn't plopped water droplets down on me sooner or later.

The fourth thing I think of, however, is the beautiful, look of the "new" metal roofs I see on a lot of the latest commercial construction—like schools, banks, government buildings, upscale malls, professional buildings, and the like. I think those roofs are gorgeous. And, so, in spite of the three previously mentioned reasons why I felt I would most likely never get a metal roof, I actually considered putting a metal roof on our barn and arena. Well, I considered it for a minute—and only from a visual beauty perspective, and clearly not from a functional or durability perspective.

LESSON #53: The point of note here is that when metal pole barn suppliers are talking about a metal roof, they are NOT talking about the kind of metal roof you see in the "new," upscale construction of our times.

The type of metal roof you see in upscale commercial construction is a very expensive type of metal roof called standing seam. The type of metal roof the pole barn suppliers are talking about in their base quotes is just a run of inexpensive, corrugated sheet metal that has little to no relationship to the beauty of standing seam metal roofing. The inexpensive copy is about as ugly as it gets. It's just a basic sheet of steel with repeating rows of ripples pressed into it.

So if you decide to go with the "standard" metal roof, you will not only potentially have noise, leaks, torn insulation and rust, you will also have abject ugliness.

Of course, you could upgrade to the beautiful and expensive standing seam metal roof, however, you will still have the noise and additional basic problems of metal roofs with which to deal.

There is another major problem with metal roofs. Metal roofs have to be insulated. If you think about barns where you have boarded, shown or visited, and you can remember looking up and seeing open batts of insulation clumped along the ceiling, you have been looking at the insulation under a metal roof. You may have also noticed that those insulation batts under the metal roofs have been ripped here and there by birds, that the insulation batts have discolored to varying shades of pink, gray, and brown from water leaks, and that the insulation batts generally lend an overall disgusting look to the barn and arena ceilings. Nevertheless, if you have a metal roof, then it must be insulated because of condensation.

The condensation occurs under metal roofs because the metal of the roof is cold, while the air within the barn or arena is both relatively warmer, as well as relatively moister than the metal of the roof. When the warm, moist air contained within the structure reaches the cold, metal ceiling (without insulation), the warm, moist air condenses on the cold metal roof and forms water droplets (also known as "rain"). To stop the formation of condensation on the ceiling, fiberglass batts of insulation are installed right up against the metal ceiling. However, if the insulation batts get ripped for any reason (like by birds pulling out the fiberglass from the batts to use for nesting material), then the rips in the insulation will allow the warm air within the structure to reach the cold air of the roof and cause the described condensation process to occur—but behind the insulation batts. Then the rips in the batting, and the subsequent condensation that results to form "rain," in turn make the tie-dyed looking water stains on the insulation batting and the occasional plop of "rain" down onto you.

One prospective builder my husband had found took him to a barn and arena he had just finished building. With great pride, the builder showed my husband how he had "solved" the metal roof insulation issue. He decided that he could "insulate" a metal roof by creating an air gap between the roof metal and sheets of plywood underneath. That means there was the steel roof, an air gap of a couple of inches, and then sheets of plywood. The builder argued that the air gap would act as a "natural insulator" for the steel roof and eliminate the need for the ugly insulation. Proud of his insightful solution to the metal roof, condensation dilemma, the builder then proceeded to pressure my husband to allow him to include the same kind of roof insulating system in our barn bid.

My husband, being an engineer, looked incredulously at the builder for a moment. Then my husband tried to discuss with the builder the fallacy of his thinking in having done his "creative" insulating. My husband tried to point out to the builder that because it would be impossible for sheets of plywood to be installed in a way to form an airtight chamber, the room air would still flow up behind the plywood as if it wasn't even there and then condense on the metal of the roof and cause "rain." Then, unfortunately, all that condensation "rain" that was created would be trapped behind the plywood which would not only leak, but eventually rot the plywood. What the builder had done in actuality was build an even more efficient cloud chamber than ordinary metal roof, fiberglass batt insulation would create. Nevertheless, the builder continued to argue his point of view, and so my husband concluded the discussion by informing the builder that there was no way we would ever consider having such a roofing system.

NOTE: When my husband tried to contact this builder about a month later, to ask when his quote would be ready for our facility, we couldn't get in touch with the builder. So, we went back to the barn where he had taken my husband—the barn with the cloud chamber ceiling--to find out that the barn owners were suing the builder for the "water" problems they were having in the arena that resulted from the way the builder had "insulated" the roof. No surprise! We subsequently learned that the builder had left town and gone out of business.

Your other option for roofing is **asphalt shingles**. Not only are asphalt shingles attractive, they are quiet, and they are insulating in themselves without the need for any additional insulation on the inside of the roof. As such, your barn and arena have clean sheets of plywood roofing on the inside for you to look up at and enjoy—and without a "rain forest" effect. In addition, the birds don't have any insulation to pick at and use for nesting materials thereby keeping your barn and arena much cleaner. Without readily available nesting materials, the birds are also not as inspired to invade your barn and nest there.

NOTE: Remember that a pole barn quote will generally only include the roof being made "ready" for asphalt shingles, and not include the actual shingles themselves. That means the pole barn builder will install a particle board roof instead of their cheap copy of a standing seam steel roof on your structures.

Be warned that if a "standard" asphalt shingle option is offered to you by your pole barn builder, that the asphalt shingles will be the cheapest, least attractive type of asphalt shingles available. Interestingly, the cost differential between "cheap" shingles and the best shingles is so minimal that there really is NO reason to consider using anything but the "best" or highest quality ("highest weight") shingles. Of course your pole barn builder would not be the one with whom to discuss the price differential of shingles because suddenly there would be a big price difference if he quotes on the upgraded shingles. You should have a separate supplier to supply and install your roof shingles under the supervision of your barn finisher builder. In this way you will avoid getting taken advantage of by the pole barn supplier and winding up with either inferior shingles or a price-gouging cost for upgrading to the best shingles.

What I mean by the "best" asphalt shingles are architectural dimensional shingles which have a layered look that gives both beauty and depth to your roof. These shingles are generally warranted to last 50 years. Architectural shingles should have a minimum weight of 300 pounds per square where one square measures 10′ x 10′.
(See Photo #55)

Be sure to install a **ridge vent** (roof ventilation ridge) along the top of each roof line of your asphalt shingled roof. This vent serves multiple purposes. First of all it allows more air flow into the barn which helps keep the overall air healthier for your horses. In the summer it allows excessive heat to escape from the barn through the roof. In the winter, it keeps the snow from continually melting off the roof and forming potentially dangerous and destructive icicles. Also by limiting the snow melt, it allows the roof snow to blanket the roof and act as an insulator for the barn. You should know that if you have a snow season where your barn will be constructed, then your pole barn supplier will calculate an estimated, average, yearly snowfall for your region or **"snow load"** so that your barn roof will be built with sufficient strength to be able to support such a snow load.

A ridge vent is preferable to installing roof pot vents because a ridge vent provides better overall ventilation, and doesn't jam with snow the way pot vents tend to do. Also, pot vents lined up across your roofing are less attractive than a subtle ridge vent.

NOTE: It is very important to insist that your pole barn supplier/builder lines the entire length of the inside of the ridge vent (on the underside of the roof) with screening to keep out all the "critters" such as nesting birds that we discussed earlier re: cupolas.

Although we have mentioned the barn aisles in passing, let's talk about the barn aisles now in more detail.

I have suggested you have nice, roomy, 12 foot wide aisles. As I mentioned earlier, a wide aisle not only allows for easy horse handling, it also allows for easy access into, and down, the aisles by your farm vehicles and equipment including the likes of your Gator and tractor and potentially a muck spreader.

But now let's consider the composition options for the aisle floors themselves.

Well, first, there is the basic dirt floor aisle. A lot of show barns I've been in have dirt floor aisles. I personally hate them. Not only are they dirty, and give a dark, damp, dreary feel to a barn, they are also always uneven and rutted. But if that's all your budget allows for, it's certainly a traditional and acceptable aisle floor for your barn. Plus, you can always either level out the dirt when it gets ruts or cover it later with another substance layer, like cement, when your budget will allow.

I'm sure you realize by now, from various references that I have made, that my barn aisles are cemented. I like cement barn aisles. I think they give structure to the barn, allow for cleanliness, and afford a very pleasant look and feel to the whole barn. However, not everyone would agree with me. Some people would argue that standing on cement in the aisles is hard on horses' legs, and is especially uncomfortable for the horses to stand on in the winter with their metal shoes. But my horses aren't standing on the cement in the aisles for any extended period of time. In fact, they only briefly walk on the cement aisles to be turned out or led into the stall mat covered grooming stall or into the wash rack. Of course, if you don't have a dedicated grooming stall and plan to use the aisle for grooming, then maybe there's a slight argument for the aisle cement being uncomfortable for the horses. However, that problem can be quickly, easily, and cheaply eliminated by simply putting down a length of rubber grooming mat in the aisle over the cement.

Other people would argue that aisle cement easily cracks and eventually breaks apart. Well, yes, that's true about cement. In fact, there will be cracks in the aisle cement almost immediately after it dries. But the length of time before cement aisles actually break apart is considerable (like 30 years maybe???) and not something I choose to worry about. Plus, there are patch systems available for cement if needed, and you can always have the aisles dug up and re-cemented if need be.

NOTE: As previously mentioned, it is important to have your cement aisles "**broom finished**" to provide a rough, non slippery surface on the top of the cement.

NOTE: If you have bare cement aisles in your barn, be careful not to play with your dogs on that surface more than occasionally. When dogs chase a ball (or other toy) on cement, their nails are quickly worn down to the quick. Depending on the dog and the starting length of the nails, just two days of ½ hour, ball throwing, play sessions can wear down the nails to the point of great pain for the dogs.

NOTE: Speaking of dogs, playing in the arena sand doesn't work out well either. After just one throw, the dog's saliva on the ball (or other toy) makes the sand cling to the object such that when the dog picks up the toy, the sand gets into the dog's mouth and subsequently gets into the dog's digestive tract. Also, when the dogs come to a "screeching halt" in the sand and/or when they pick up the toy, they are inclined to get sand in their eyes, ears, snout and mouth—none of which is any good for them.

The dog on cement problem brings to mind a costly, but nice way to finish barn aisles. Cemented aisles can be covered with attractive, cushioning, rubber paver "bricks". There are many online dealers as well as online installation instructions and prices to give you a feel for whether or not your budget will allow for rubber paver brick aisles.

There are other aisle floor options, some of which are geographical and climate specific. For example, barns in warm climates, near the ocean, frequently have crushed shell coverings over their dirt floors. So there are lots of different aisle floor options, but, again, I personally prefer cement.

Let's now "look up" from the aisle floors and consider the **aisle walls** in your barn. Right now you have just the basics on your aisle walls--the stall fronts where there are stalls, and the interior side of the pole barn metal that covers the exterior walls of your barn.

First, each and every one of the stalls must be lined with #2 southern pine, tongue and groove wood at a minimum height of 8 feet. Each stall will have a stall front facing the aisle, then have three other sides composed of #2 southern pine, tongue and groove at 8 feet of height. Those three walls include the two walls, one on either side of the stall, that separate the stalls, one from the other, as well as a "back" wall to the stall, that also lines the interior of the exterior metal wall of the barn. **(See Photo #65)**

NOTE: The back wall in my stalls is actually 8 and ½ feet high.

Now what about the rest of those metal shell, exterior walls in your barn?

I would strongly urge you to include the money in your budget to have your barn builder/finisher line the interior metal wall of every aisle wall in your barn with the same #2 southern pine, tongue and groove as your stalls, and to the full height of the wall from the aisle floor to the roof edge or at least to a height of 8 feet. Not only does this wood lining give a beautiful, flowing, warm look to your barn, it also gives strength, and some insulating factor, to your barn. I would also suggest that you fully line your grooming stall as if it will be a future stall with walls lined at an 8 foot minimum height with #2 southern pine, tongue and groove. As I have previously mentioned, in the tack room, lining all walls, floor to ceiling, with the tongue and groove pine is both functional and attractive. Finishing off the tack room ceiling with T111 wall board (which has a bead board look) blends in nicely with the look of the tongue and groove on the side walls. Again, with the entire tack room lined with wood, you can easily nail up saddle racks, bridle racks, blanket racks and anything else you might want to hang on the wall!

There are other types of wood siding than the #2 southern pine, tongue and groove available for lining your barn walls. However, as with everything else, the other woods that you can consider using might be more expensive.

NOTE: The bathroom walls and the walls in your office/lounge/viewing area can be lined with the tongue and groove pine as well. However, a less expensive option is to use sheet rock and wood trim which gives those rooms more of a "house like" feel which works well in those particular areas as I previously mentioned. In my opinion, tongue and groove in the bathroom and office/lounge/viewing area gives too "heavy" a look in those types of rooms.

Chapter Seven

Arena Design and Construction

Photo #66: Construction of Our Arena

Now it's time to talk about **arenas**. We previously mentioned arenas regarding the limitations of arena width by prefabricated trusses as well as the standard sizes for dressage and jumping arenas.

Just for a quick point of review, the standard size for a competition arena is 100 feet x 200 feet. A jumping arena (not for competition) is generally 80 feet by 200 feet. The standard sizes for dressage arenas are: (1) 66 x 132 feet for a small arena; and (2) 66 x 198 feet for a large dressage arena. The widest, prefabricated trusses that you can buy are 72 feet with the minimally attractive 3.5:12 roof pitch. If you want a wider arena with a minimum 3.5:12 or higher pitch, then you will have to have your trusses built onsite which could be prohibitively costly depending on your budget. A 72 foot wide by 136 foot long arena is fine for jumping three practice lines and can make use of the center line to jump twice to form a full, four line hunter course. A 72′ x 136′ arena will also fit a small dressage arena within it.

Those facts aside, I would like to make a general comment about arenas. It is important to have an arena whether you are in a place where winter does or does not happen, or if you are in a place where it rains most of the time, or if you are in a place where the sun shines most of the time. Arenas are unequivocally the horse owner's best friend. Do whatever is necessary to have enough money to build an arena along with your barn, even if it's smaller than your ideal sized arena.

NOTE: Remember that if you plan your building site for future arena expansion—in other words, you have made sure that you didn't build your arena right on the setback line of your property--and you have aligned your arena in such a way as to be able to make it larger in the future, then you can always expand your arena if it proves to be too small.

When you have an arena, you always have a place to ride as well as a place to do **turnouts**. You will find that having the ability to do turnouts in an indoor arena is endlessly reassuring. No matter what the weather conditions are or what your condition is to be able to ride, you will at least know that you will be able to turn out your horse to get his daily exercise. I really cannot overemphasize the relief that knowledge will offer you in your horse management. Plus, horses love the security of the arena. They think of the arena as an extension of their stalls, they see the arena as a happy and safe play place where they can expend any and all excess energy, and then calmly wait for you to collect them at the gate when they are done playing.

That said, now let's talk about the other features to consider when you are planning your arena.

If you have jumping horses and plan to use your arena to jump, then the height of your arena exterior sidewall must be at least 16 feet in order to provide a minimum of 14 feet of unobstructed clearance (for jumping) between the arena footing and the lowest point of the roof trusses plus the allowance for about 2 feet of clearance for lighting fixtures that hang down from the ceiling trusses (14 feet for jumping + 2 feet for lighting fixtures = 16 foot exterior side walls). Combined with the 3.5:12 roof pitch, this 16 foot exterior wall height also gives a wide open, inspirational feel to the arena.

Photo #67: Arena Interior, Truss Ceiling, Low Bay Lights, Translucent Panels, Kick Wall, and Open Slider Doors with Kick Wall Swing Gate

In order for your arena to look proportional to your barn from the exterior due to the difference in roof pitch between the barn and the arena, the exterior side walls of the arena should be proportionately taller than the exterior side walls of the barn even if you don't need a full 16 foot exterior wall height for jumping horses. Remember that the roof pitch is different on the barn and the arena because the roof pitch on the arena was compromised by the width limits of prefabricated roof trusses. With proportionately taller arena walls than those of the barn, even though the roof pitch of the arena is lower than that of the

Photo #68: Relative Roof Pitches of Barn versus Arena

barn, the profile of the whole project will look correct and attractive. But, again, this is a money issue, and if your budget allows, both structures can have the same roof pitch and then the exterior side wall height is not necessarily an issue.

Photo #69: Exterior View, Arena Wall Translucent Panels

Once again, you should have a ridge vent running along the entire length of the arena roof. Also, again, the entire roof ridge vent should be lined with screening (on the underside) to keep out any unwanted "visitors" to your arena.

Translucent panels should be installed along the exterior walls of your arena to allow in natural light during the day. Lining the entire length of both of the "long" exterior walls of the arena with translucent panels makes for a perfectly lit (during the daytime!) arena. The "clear" translucent, plastic panels (as versus opaque panels) allow in the best light. These panels should be 6 feet high and should be installed 10 feet up the wall of the arena from the footing. If your arena has walls higher than 16 feet, then your metal exterior walls would begin again above the 6 foot light panel. For example, a 20 foot exterior side wall would have 10 feet of metal siding, then 6 feet of translucent light panel, followed by 4 feet of metal siding.

You must next plan for entry into your arena. An arena needs to have an entry gate big enough for you and your horse to have easy access into the arena from the barn.

Photo #70: Arena Entry Gate from Horse Barn Connector Aisle

The arena end doors also need to be big enough for easy access into the arena for the tractor and the drag from outside the arena, as well as for at least a small sized dump truck that will need to access your arena, at a minimum, during construction to deliver your arena footing.

When considering the entry doors into your arena, you should also consider air flow. Strategically placed arena doors can create a wind tunnel effect for air to flow either from the barn and out of the arena or vice versa. A wind tunnel effect can be a wonderful asset—especially if you live in a year round warm climate—or it can be a nightmare if you live where there are long, cold winters. Of course I don't mean to indicate that you would have the arena doors open in the winter. What I am talking about is the reality that there will be air leaks even when the arena doors are closed which will cause wind tunnel effects, from minimal to moderate, and during the few times you will have the doors open even when it is winter (like when you bring in the Gator and arena drag to drag the arena). There will also be temperature differentials that occur between the relatively warmer barn (due to the horses producing body heat) versus the cold, empty arena—especially at night in the winter—that will create significant, pulling air flow.

Photo #71: Arena Slider End Door

Photo #72: Arena Kick Wall with Cap

Your arena will also need a kick wall that runs around the interior of the entire structure. The kick wall should have a minimum height of 4 feet and be minimally constructed from 5/8" thick plywood. Remember, horses kick! They are especially inclined to kick when they are playing in the arena. If you don't line your arena with a wood kick wall, you will have huge dents in your steel metal siding rather than kicked in holes in easily replaceable plywood. Your kick wall should also have a wood "cap" of 2" x 6" strips of lumber running along the top of it. Although the cap won't keep mice out of the arena as such, it will certainly discourage them from coming up into the arena. The cap also keeps down the accumulation of debris between the arena exterior wall and the kick wall.

NOTE: The kick wall must continue across the full expanse of the backside of any entry doors into the arena with kick wall gates. The kick wall gates need to swing open to allow in the tractors and trucks, and then be able to securely close with gate closure hardware to protect the metal of the entry doors. These gates will also allow you to open the entry doors in the warm months for air circulation, while containing the horses behind the kick wall gates in the arena—good for air circulation, turnouts in the arena, and riding in the arena in the summer.

Photo #73: Opened Arena Slider Showing Kick Wall Swing Gates

Okay, with those basics said, let's talk about the details. Once again I will present the information based on my original thinking and then note how things turned out.

In my basic plan, you will note that I have a nice, 12 foot wide, connector aisle running from the barn aisle, down roughly into the middle of the length of the arena. Again, I will mention that the width of this aisle is not only pleasant, but very functional. The 12 foot width gives plenty of room to stow show trunks and blanket storage boxes, as well as to accommodate sofas and chairs placed along both walls, plus plenty of room for me and my horse to walk together down the aisle.

Photo #74: Alex Enjoying the View Thru the Opened Arena End Door

There were two other reasons why I wanted to have a wide connector aisle. One was that I had originally planned to use the aisle for temporary hay storage by parking a hay wagon in the aisle. The other reason was that I wanted a wide opening from the connector aisle into the barn so that I could potentially park an ATV with an arena drag in the aisle and/or park the Gator in the connector aisle and be able to easily drive those vehicles into the arena from the barn.

NOTE: The floor to ceiling height (to the bottom of the trusses) in the connector aisle is 11 feet. I made the corridor ceiling high both for attractiveness of roof lines and to allow head room should a horse rear up in the aisle. I also added translucent light panels along the top of the exterior wall of the connector aisle. These panels are 2 feet in height and begin at 9 feet above the aisle floor. They give the connector aisle pleasant natural lighting during the daytime hours.

Photo #75: Connector Aisle with Alex at Arena Entry Gate, Connector Aisle Truss Ceiling, Translucent Panels on Exterior Wall, and Spray Foam Insulation covering Cross Over Heating Tubes in Aisle Bridge

Well, having enough room to drive the Gator or an ATV into the arena from the connector aisle didn't happen for me because when I arrived at the barn one day, I found that the builder had built half walls across the connector aisle opening into the arena with a 4 foot wide, kick wall, swing gate in the middle of the stationary half walls. That meant that I could enter the arena with my horse, but I couldn't enter the arena with an ATV or Gator from the corridor aisle because the half walls had essentially narrowed the entrance to the arena to a width of four feet.

During the planning stage, I also instinctively knew that I would need some kind of fully closing door at the end of that connector aisle where you entered the arena, in addition to a man/horse gate through which to enter for riding. I had planned to have a garage door installed there with an automatic opener. The purpose of the garage door would be to enable me to completely close off the arena from the barn. Not only did I feel this would keep the barn warmer in winter, but I felt it would keep down any "critter" traffic that might occur between the arena and the barn.

Well, I was right about the need for a garage door in that at night, during the winter, the arena does indeed suck out all the warm air from the barn making the barn colder than it should be during the night. Nevertheless, my plan to have a garage door to block the air, heat exchange from the barn to the arena, also didn't happen for me. Our builder, the one who left town, didn't bother to fully plan for the installation of the garage door either which meant that installing one "after the fact" became a cumbersome kluge that would result in a lot of cost as well as the additional, deal breaking realization that the "ugly" side of the garage door would be the one facing into the barn—that is, assuming that the installation was really doable at all.

However, if you aren't bothered by having the "ugly side" of the garage door facing you everyday at the end of your barn aisle, then you can plan the installation of one from the beginning of your planning which will at least be cost effective as versus my situation. You will need to consult with a garage door supplier/installer in the planning stage of your project to find out "exactly" where he needs power sources installed for the automatic openers, how high the header above the entry door needs to be for him to install the garage door track, how high he can possibly install the track to give your horse the most possible height clearance to pass under it, and how much wall width he needs on either side of the entry for the installation of the garage door.

You might be thinking to yourself, "Well, why not just put the track on the arena wall so that the 'pretty side' of the garage door faces into the barn aisle?" The answer is because: (1) the garage door tracks would be sticking out from your arena wall and could therefore possibly injure you and your horse while riding or your horse when playing in the arena; and (2) because when the garage door is opened, it would open up onto a track with the door itself suspended and floating (so to speak) in mid air over your riding area which would be dangerous, ugly, and potentially spooky.

Then you might also ask if there are garage door tracks that can run straight up the wall? Yes, there are. However, once again, you would not want to install the door on the arena side of the wall because of the danger of the track edges sticking out into the arena, even if the door is going to open straight up. But you could have everything work, including having the pretty side of the garage door facing you, if you install the garage door on the barn side of the wall and have it open straight up. But that can only happen if you have a really high ceiling in your barn corridor to allow the door to open straight up the wall. Then the problem that can arise with a really high barn corridor ceiling would be that such a high ceiling line could throw off the lines of your barn in relation to the arena as well as increase your costs. Nevertheless, if you plan for enough height in your barn aisle and use rafters rather than trusses in the connector aisle, then you could install a great door that opens straight up and has the "pretty side" facing into the barn.

Note: Finding a cooperative garage door installer who will customize your installation may not be easy, and customization always exponentially increases your costs.

There are other solutions to the air flow problem between the barn and the arena that could easily be solved by planning ahead. The most logical "plan ahead" solution would be to make the entrance opening into the arena the width of two standard doors. The reason I have had such a cost prohibitive problem with closing the air gap is that the builder constructed the opening with an odd sized cut out, and with little wall space on either side of that opening. The open area above the kick wall and gate of my entry into the arena is a non standard opening of 5 feet high by 9 feet wide. The width of the opening only leaves me 1 and ½ feet of wall space on either side of the opening. The additional problem is that I have an exit door on the exterior wall of the connector corridor that is just two feet back from that entry gate.

The first problem is that standard door widths are 48". Two standard doors equal 8 feet of width. My opening is 9 feet wide and that means I would have to get custom doors made to close the gap or frame in the gap. Buying custom doors each at 4 and ½ feet of width would mean I would have to pay about four times the price of standard doors. Even if I got the custom doors, I still wouldn't have enough wall space for the doors to open flush against the barn wall—and of course I wouldn't want to have doors open from the arena side of the wall and lay against the interior wall of the arena—either option again being potentially dangerous to horses and riders. Then, too, the opened door in the barn, even if I felt it would be okay non-flush against the wall when opened, would open back into the exterior door in my connector aisle and would block its use.

To solve the problem, I have also considered custom, wood, Dutch windows for the opening. But the issue with the fold back problem of Dutch windows is really the same as the door problem. Folding shutters are another option in that they could fold back into each other, but shutters allow air flow through them making them ineffective in solving this problem and they are also very expensive.

I then considered track shades for the 5 x 9 foot opening. For a brief moment, that seemed like the perfect, cost effective solution to the problem. Then my husband pointed out that the shades wouldn't be strong enough against even the most minimal pull (wind) created by the heat exchange between the barn and the arena and would actually cause the shade to turn into a billowing sail between the two structures, rip out, and therefore not work.

I subsequently found a business online that makes "wind resistant shades and tracks." These are the kind of heavy canvas shades you would see at a beachside or dockside restaurant. Of course, what immediately comes to mind—if you've ever eaten at such a facility—is that billowing sail effect my husband mentioned. But at least with a heavy fabric and sturdy track I'd have more of a chance of the shade not ripping out. That seemed like a good option until I found you needed a commercial restaurant-sized budget to afford such a shade and track!

I've also considered putting in some kind of Plexiglas solution—also very expensive. Lately I've considered building up the walls on either side of the actual swing gate to full height, and then just installing a Dutch door top above the entry swing gate into the arena that can lay back against the built up wall next to it. Of course, that would mean limiting the view into the ring from the aisle which would make the look of the entry, and the whole barn corridor, much less appealing.

The current solution is the same one that solves the issue of whether or not to heat the barn. I just put an extra blanket on the horses on the cold nights! So far that's been the easiest and cheapest solution.

The reason I went through all my thinking in how I have tried to solve my particular problem is to show you all the kinds of "enclosure" options that are available to your thinking when you are planning ahead as to how to separate the air flow between your barn and your arena.

<u>**NOTE:**</u> Remember that if you are going to be a commercial facility from the start of your project you will probably be required to install a very expensive, fire proof door between the barn and the arena which the local building inspector would specify.

Well, anyway, you have your entrance gate from the barn corridor into the arena. Now you will also need to determine which way you want that entrance gate to swing—either into the arena or into the aisle. I would recommend having it swing into the arena. Yes, you will have to look into the arena and make sure there isn't any horse traffic in the way before you open the gate. However, that is better than opening the gate into the aisle and having your horse bolt ahead of you into the arena and crash into traffic you haven't seen before opening the gate!

Photo #76: LeTigre Standing on Arena Entry Gate Hardware Protective Wood

The gate also needs to have a heavy duty closure on it that automatically hooks when you close it like your pasture gates. Around any such protruding gate closure hardware, should be some kind of protective wood on the arena side of the gate so that the horses can't rub against, or get caught on, the gate closing hardware and injure themselves. My horses regularly rub against the protective wood around the gate hardware when they are standing by the gate waiting for me to come get them. At first, when the protective wood was just nailed onto the gate, the horses were able to rub the whole piece of wood right off the gate. Therefore, nailing in the protective wood around the hardware is not sufficient. The protective wood must be bolted into place with BIG bolts. Also in the process of rubbing, sure enough, even with the protective wood, the horses can manage to access just enough of the entry gate hardware to open the gate!

The times when my horses have managed to open the arena entry gate, it's been rather funny to see them standing there at the opened gate looking down the aisle at me as if to say, "Look at what I did! I opened the gate! Aren't you proud of me?" But then their next immediate expressions seem to say, "But to tell you the truth, I'm still not sure somehow what I'm supposed to do now that I've opened the gate. Am I supposed to stand here and wait for you to come get me or am I supposed to come out into the barn aisle and get you?" Fortunately, to date, I've gotten to the horses at the gate before they've used their horse brains and drawn a horse logical conclusion that maybe they should make a "fast dash down the aisle to their stalls or out the barn door!"

Photo #77: Angled View of Arena Entry Gate Hardware Protective Wood & Bungee

What I've done now to keep the horses from accidentally opening the gate, is to attach a bungee cord between big eye bolts my husband attached on either side of the gate and kick wall (one eye bolt on the arena side of the gate; the other eye bolt on the barn aisle half wall) that runs across the hardware and keeps the gate closed.

Photo #78: Slider Door Locks

So now let's consider how many big entrance doors you should have into the arena from the three remaining walls—those walls that are NOT the aisle connector wall between the barn and the arena and where the horse/human entry gate is located.

NOTE: Generally speaking all arena doors, hay barn doors, and shavings barn doors are just made by using sheets of the exterior metal. As such, these doors are relatively flimsy and prone to being affected by the wind—from rattling in place to actually blowing off. Having center stabilizer bars in the ground (installed below the freeze line) are very important for this type of door.

NOTE: Be sure all your slider end doors, and all your slider entry doors, that are more than 8 feet in width, are center opening sliders to reduce the square footage of each door you will be handling. Also be sure that every slider door (as well as every man door) has a locking system included so that you can secure all the doors on all the barn buildings.

Photo #79: Slider Door Lock

In my opinion, you should have at least two opposing doors in the arena. First of all, this will give you nice air flow for indoor riding in the summer and for indoor, summer turnouts. But it will also allow the construction trucks to drive in one door and out the other. The easier you make it for the construction vehicles to access and then egress your facility, the less risk there will be for damage to your facility! Having the two opposing doors makes it easier for your tractor, Gator, and/or ATV to access and egress the arena as well. Plus, by having two doors, you will have the option of leaving and entering by the one door if the other door should become dysfunctional for some reason—like being frozen shut!

NOTE: I installed two opposing doors in my arena, each opening being 10 feet high by 12 feet wide which is the same size as the end doors in my barn. This size opening allows in pickup truck sized vehicles, normal sized tractors, and small dump trucks. However, my doors are too small to allow in big trucks or construction vehicles. If you feel that you will need to have big trucks entering your arena, then at least one of your entry doors into your arena should be approximately 14 feet high by 15 feet wide. However, it is important to remember that big doors are harder to handle, rattle more in the wind, AND the arena kick wall gate required to span a 15 foot wide opening will be very wide and hard to keep from sagging. Because we did not have a 14 x 15 foot opening, our arena footing materials were delivered outside the arena in piles. Then smaller construction vehicles were used to transfer loads of the footing materials into our arena as seen in **Photo #71.**

Sometimes I wish that I had a third door on the arena that would line up with the corridor aisle and the front door of the barn. I think that it would make a nice breeze through the barn in the summer as well as allowing more air flow in the arena while riding and turning out indoors during the summer. However, as we've already discussed, there already is an air pull without such a third door on the arena that occurs between the barn and the arena that makes the barn colder than I would like it to be in the winter. Also, when riding in the summer, the open arena doors are always a point of potential spook. Having two open arena doors while riding indoors in the summer is enough of a challenge. Having a third arena door open while riding and increasing the potential spook factor by 1/3, might be too much. That's a decision that can only be determined by you and your comfort factor with an opening to ride by on each of the arena's four walls.

NOTE: Don't forget that for every arena door you have, you will need to have an arena kick wall, swing gate to swing open to allow in vehicles and equipment, and then to swing back and lock to keep your horses from running out of the arena or from kicking off the metal arena doors.

Remember that little southern belle barn designer? She had another suggestion for arena air flow that she was almost obnoxious about wanting us to install. Even though this idea didn't work for me, I am including it for you to consider because it might appeal to you—especially if you live in a warm climate. The option she presented was to have the entire length of both your arena long walls have a continuous line of opening garage doors. In this way, in the warm months, it would be possible to totally open up the entire length of both your arena long walls. She pointed out that if you installed these fully opening walls, that there would be no need to have an outdoor riding ring, because your indoor ring would become an "outdoor" ring when you opened all the garage doors.

This was actually an attractive idea at first until we started pricing it out and really thinking about the overall functionality of it. Other than the cost issue which was huge, the next major problem was that there would be a spider web of garage door tracks running mid air all along the arena with mid air, "hanging" garage doors when they were open--neither situation being good for jumping horses. Then, too, once you cut holes in your arena walls for garage doors, you would be letting in too much cold air from drafts in the up north winters—even when the garage doors were closed. You would also get lots and lots of garage door rattling when the wind blew and who wants to try and ride a horse when all the walls are rattling and squealing in the wind? Then, the chance of water leaks would increase exponentially as well as the probability of costly and inconvenient repairs to broken, faulty garage doors.

The bottom line is that this might be good for arenas in the Pacific Northwest where it rains a good deal of the time or in the South or Southwest where the heat gets oppressive in the summer, and having a covered, open area in which to ride would be desirable. Nevertheless, it doesn't really work that great for a four season climate with cold, windy winters, and relatively cool summers.

Another door consideration for your arena would be whether or not you want to have a man door on one of the exterior arena walls for easy access to walk directly into the arena without having to go through the barn or open a huge arena end door. I didn't really have a need for a man door and didn't put one in. I also personally feel that man doors that can't be easily observed are points of entry for potential trespassers. Therefore, I am not inclined to put any extra man doors beyond the essential ones. Also, having a man door suddenly open into the arena while someone is riding could easily cause a horse to spook and cause an injury.

The next consideration for your **arena** is the **footing**. Arena footing is a major issue. There are a lot of alternatives. For example, there is a footing additive that has varying degrees of shredded, old rubber tires mixed in it. Supposedly this footing has better resiliency, is less compacting, reduces dust, and lasts forever. And that may all be true. However, when I saw all the available samples I had to select from, I just got plain confused as to which one would be appropriate for jumping horses, and the sales representative didn't have any great suggestions to help me solve my confusion. I also felt that having ground up rubber tires could be really smelly—you know how stinky tires smell. In addition, I saw the "lasting forever" feature as ultimately being a problem, in my opinion. Even though the rubber in the footing may last forever, I felt that the footing would eventually get so dirty or bacterial/fungal ridden that it would need to be replaced. Then, where are you going to dispose of used rubber tire footing? That could become a huge issue. So, I decided to cross this type of footing additive off my list.

Another footing I've been exposed to was at a major "A" show that is held indoors each year. The footing they used was shredded bark which is also called mulch. And, yes, if you're thinking "Isn't that the same stuff they use in gardens?" that as far as I could see, it was the same stuff. I was actually rather amazed when I first saw that mulch was being used as footing. The first thing that came to my mind was that it would be a surface that would seem to quickly and easily breed a lot of bacterial and fungal problems. Of course the mulch was probably "treated" with something to minimize such problems, but it just looked a little TOO organic to me and somehow I just didn't feel comfortable about it and decided not to use this footing either.

The other footing I've had personal experience with is sand mixed with fiberglass. Frankly I couldn't believe my eyes when I saw the fiberglass insulation mixed in with the arena sand. In fact, my eyes were the first thing I thought about should I fall off my horse considering I once had the misfortune of having a piece of fiberglass fall into my eye while helping my husband lift fiberglass insulation batts up a ladder into our house attic. Let me tell you that it hurts! Also let me remind you that the reason it is called fiber-"glass" is because there are little pieces of glass in it! Now imagine turning your horse out into your arena to roll in this type of footing. Just think of the potential pain and suffering you would cause your horse because sooner or later he would surely get pieces of the fiberglass in his eyes, as well as in his large, and sensitive nostrils and muzzle.

You will find that most riding rings just have some sand over the natural substrate of the land. That seems to work okay, but usually horses—especially jumping horses—start getting tendon and ligament problems over time from the inadequate stability of the footing and the lack of sand depth for cushioning.

Proper arena "sand" footing has three layers consisting of three different materials. The depth of the various layers will be determined by whether you and your horse will be jumping in the arena or just doing flat work. Proper arena "sand" footing is what I used for our arena.

First you should select a location for your indoor riding arena that will have the best drainage. Hopefully you have sited your indoor arena on the high point of your property.

NOTE: If you are going to put in an **outside riding ring**, it is critically important—due to rainfall and run off—that the ring be situated on a high point on your land, and that the sub base of the ring be "crowned" for proper drainage (crowned meaning that the sub base is higher in the middle and then slopes out to the edges).

Next, for the "sand" footing in your indoor arena, your excavator must scrape off all the topsoil down to the natural substrate of the land. Once you have the substrate exposed, you will need to determine if that substrate consists of sufficiently stable material to be used as your "first" or subbase layer. Your excavator should be able to make the determination of whether your substrate is stable enough to use as your first layer. However, if he cannot determine that for you or if the excavator tells you your substrate is unstable, then you can hire a soils engineer as a consultant who can tell you if your substrate is stable or what you need to add to the substrate to make it stable. If you live where the climate has deep freezes and thaws, then your first layer should NOT be the natural substrate, but should consist of stone. A stone subbase should consist of 4 to 8 inches of **22AA limestone gravel aggregate with stone dust**.

Next, whether you are using the natural substrate or the 22AA limestone gravel aggregate as your sub base, you will need the excavator to compact the sub base by using a 20 ton roller (minimum 8 ton roller). The sub base must be compacted to 92% or better compaction.

Next is the base layer. The base layer is made using limestone screenings, also known as **stone dust**. You will need 6 inches of stone dust if your arena will only be used for flatting. You will need 8 inches of stone dust if your arena will be used for jumping. The stone dust is then wetted and rolled and compressed to 4 inches for flat work arenas, and to 6 inches for jumping arenas. The calcium contained in the limestone, when combined with water and then rolled, will cause the stone dust to congeal into a stable base. **This layer of stone dust is essential to isolate and prevent the gravel in the sub base from moving upward into the top layer usually built with sand.**

Last is the cushion layer or top layer. The top layer is made using **ASTM standard C-33 sand**. It is very important to only use this type of sand because it has a minimal amount of fine, dusty particles contained in it. Other sand is very dusty and will cause dust to billow everywhere including up your horse's nostrils and into his lungs. Arenas used only for flat work should have a top layer of 2½ to 3 inches of sand. Deeper sand of 4 to 6 inches is needed for jumping arenas. However, you do not ever want to have sand deeper than 6 inches in that it will be too deep and cause the horses to get "stuck" in the sand and suffer tendon and ligament problems. In addition, it is critically important to know that the deeper jumping sand of 4 to 6 inches must be kept wet and dragged to keep it firm and level, otherwise the horses will suffer the same kind of tendon and ligament problems from the sand being too deep.

The importance of wetting and leveling the sand for stability can best be demonstrated by thinking of sand at the beach. When you walk across the beach between the parking area and the water's edge, the deep, dry sand, is very hard to walk in. However, when you reach the sand at the water's edge, where the sand is wet and flat from the water, walking becomes totally easy, firm, and stable. This is the same principle that applies for the need to keep the arena sand wet!

Regarding the three layers of materials required for proper arena footing, I coincidentally saw a show on television the other night that said the reason why Roman roads have lasted all these centuries is because they are composed of 3 layers of different materials. Just thought I'd mention it!

Well, now you have your footing for your arena and you can feel relatively confident that your horses are not going to suffer any leg damage from inadequate footing if you've done the proper, three layer, arena "sand" procedure. I remember the first time I walked across the footing in our arena and found it to be firm and stable. Knowing now what good footing felt like, I finally understood just how bad it was for horses when they had to flat, and especially when they had to jump, on just a thin layer of sand over dirt.

Photo #80: Gator Pulling Arena Rake

So now that you have your arena footing in, and know that you have to keep it wet and level, the next thing you might find yourself wondering is just exactly how you are going to keep your arena sand moist and level.

Keeping the footing level is relatively simple. You need to have an arena rake or, at a minimum, an arena drag. The rake or the drag can be pulled behind an ATV, a Gator or a tractor. The difference between an arena rake and an arena drag is that an arena drag doesn't break up the sand and then smooth it out. An arena drag only levels the top layer of the arena sand. However, because a drag is cheap, small, and easy to pull, it is a popular way to level arena footing.

Arena rakes, on the other hand, both break up the arena footing and then level it. Until recently, however, arena rakes were large and heavy, and could only be pulled by a tractor. They were also relatively hard to handle and control. But the good news is that a new, smaller arena rake is now available with retractable wheels that can be pulled by a Gator or ATV, is easy to attach to a hitch, and is relatively easy to use and control.

I do, however, want to point out to you the one mistake we made with our arena rake. We ordered the narrower version of the small arena rake which meant that the width of the rake is the same size as the width of our Gator. As such, the rake can't get right up next to the arena wall when it is attached to the Gator, and is therefore limited in its ability to rake along the edge of the arena walls and deep into the corners of the arena. So, be warned to make sure that the rake you purchase is wider than whatever vehicle will be pulling it!

As mentioned, the second issue with arena footing is concerned with keeping the arena sand moist. When I first heard people at the various barns we visited talking about keeping the arena footing moist, it didn't really capture my attention or interest as such. After all, the footing in the indoor arena where I had ridden for so many years was always moist. Plus, I never saw them water the arena or talk about arena moisture or do any anything about arena moisture. So every time I heard the subject of arena moisture come up, my eyes just glazed over, and my mind went into la-la land.

Well, oops--I should have been paying more attention!

The irony is that when you correctly install your arena footing, you will then have the "problem" of keeping the sand moist. It turns out the reason why the arena at the barn where I boarded and rode for all those years had moist sand even though the footing was poor, was that the sand was sitting on a layer of dirt with an incredibly high water table right under it. The water table was so high in fact that in the spring when the frost would give way to the warming temperatures of the air, the cement aisles in the barn would become sopping wet. The barn would look like there had been some kind of a rain event in the aisles inside the barn—it was actually that wet. Of course I never put two and two together back then, but I now know why arena sand moisture was never a problem at that barn—and why chronic leg injuries were a continual problem for the horses.

But, I am going to assume that you will be installing proper footing in your arena so that your horse's legs will stay sound and healthy, and, that, therefore, you will have to think about how you will be keeping your arena sand moist.

At the same barn we toured where the owner had found those "wonderful" European windows, she also had a really spiffy system installed in her barn for keeping her sand wet. The system is called a Eurosprinkler system. The Eurosprinkler has specialized sprinkler heads that hang down from the arena ceiling which are fed water through a specialized pump tank, whereupon excess water is drained back into the tank by gravity feed when the sprinkling is finished. The tank must be installed in a heated room. Every part of the tubing feed system to the sprinkler heads must be installed perfectly correctly to allow for gravity feed return drainage or else any water left remaining in the tubes will freeze in the tubing during the winter months. There is specialized installation, specialized equipment and control panels, and extensive plumbing required for the Eurosprinkler system.

Although it's a great system, by the time you pay for everything, it is really pricey (about $10,000), and there's room for a lot of things to go wrong. The system also has only one supplier that we could find in the entire United States at the time we built. Considering that anything made up of parts will eventually break and need parts and service, the limited supplier didn't make me feel hopeful about parts and service happening, and added to my decision to consider other options. Plus, considering how quickly water droplets freeze on the cement floor of the wash rack in winter when I empty out the water bowls for cleaning, the chances of having water freeze in the tubing of the Eurosprinkler system during back flow and consequently rupturing the tubes, seemed to me to be quite high.

Of course, the cheapest way to water the arena is to drag a hose out into the arena and stand there and spray water on it. However, the problem with this basic method is that it takes up a couple hours of your very limited time.

The next cheapest thing to do is to buy a circulating commercial grade lawn sprinkler and place it in the arena, moving it as each zone becomes sufficiently moist.

Okay, moving a sprinkler around an arena sounds easy enough to do, doesn't it? And actually it is easy. Sure, it's a little messy in that the wet sand sticks to the hose and the sprinkler, and then of course to you. But, incredibly, even this simple system isn't as fool proof as it seems it should be. Why? Well, the first problem is that it takes more than an hour per zone to thoroughly wet the sand. So, because you will probably find that you have better things to do with your time than to stand there and watch a sprinkler spin around for an hour before you have to move it again, there's a tendency to walk off and do something else during that hour. What could possibly go wrong when you walk off during that hour and then get absorbed in another project? You guessed it—you can totally lose track of the time and maybe even totally forget you even set out a sprinkler. And what happens when you forget about the time and/or the sprinkler in any one of the zones you were watering? That's right! You wind up with quite a sizeable lake in your arena and some really BAD footing issues—for days!

There are heavy-duty "walking" sprinklers available for arenas, but once again, they are pricey. I don't have any personal experience with any of the big arena "walking" sprinklers. However, I have had walking lawn sprinklers, and frankly, they are not my favorite invention. In fact, I've had enough negative experiences with little lawn sprinklers to make me think that dealing with a big arena, walking sprinkler would just wind up being an even bigger negative. But, the fact is that I don't know for sure, and so I will leave that research and decision making for you.

NOTE: In any case with watering your arena footing, you should not let the water come in contact with the arena walls. Remember that most unfiltered water contains a lot of iron. The iron in the water will eventually discolor your arena walls to an ugly shade of brown if you allow the water to contact your walls while sprinkling.

But backing up for a minute, the next cheapest thing you can do that fits in between the low cost of the rotating lawn sprinkler, and the expensive, large arena, walking sprinkler, is the option of "brining" your arena sand. I had forgotten all about this option as well as that the farm where I rode used to brine their outdoor arena sand at the beginning of every summer outdoor riding season. I was reminded of this option one day when the mill store people mentioned it to me in response to my telling them how time consuming and labor intensive I found it to keep the arena sand moist. Although they didn't recommend "brining" the arena as such (which involves hiring a truck to come in with liquid salted water to spray on the arena sand), they did recommend "salting" the arena sand using bagged Calcium Chloride. I was summarily excited when the mill people reminded me that salting the arena sand was an option for moisture retention. I thought that by salting the arena sand, it might afford me some relief from the constant need for watering. I read online that the calcium chloride salt would retain the moisture in the sand for a long period of time (about 6 months) once I watered it.

So I immediately ordered more than a hundred dollars of calcium chloride from the mill to spread in the arena. Although the mill sold the calcium chloride (CaCl) to me, no one at the mill seemed to know exactly how you were supposed to apply the calcium chloride to the arena sand. But I wasn't worried. I just assumed there would be application instructions on the bags. However, when the bags arrived, I found that there weren't any application instructions on the bags.

So, when we got the bags of calcium chloride and there weren't any instructions as to how to apply it, we just called the supplier and asked them how to do it. But, it turned out that the supplier didn't know how to install the calcium chloride on the arena sand either.

So although we couldn't find anyone who could tell us how to put the CaCl into the arena sand, the one thing we did find out--because it was made very clear from the information that was in fact boldly printed on the CaCl bag--was that calcium chloride was an extremely caustic chemical. In fact, it was not only extremely caustic in itself, but when water was added to CaCL, an exothermal chemical reaction occurred that produced a dangerously large amount of heat that would burn you.

Well, by the time I got finished reading about all the dangers involved with using CaCl, I told my husband that I thought it would be a good idea for him to consider spreading the CaCl on the driveways to help keep down the dust out there rather than spreading any of the CaCl in the arena. In other words, there was no way I was going to put such a caustic and dangerous chemical into my arena where my horses regularly rolled in the sand on turnouts and would therefore unavoidably get this caustic chemical in their eyes—not to mention all over their coats. So, the point here is **PLEASE, DO NOT USE** any kind of **chemical "salt" additive** in your arena sand whether it arrives in bags or in a brining tank truck! By the way, I should note that the Calcium Chloride on our driveway has done an excellent job of minimizing the driveway dust!

I am confident that you can come up with some of your own creative answers as to how to keep your arena sand moist with such things as lawn irrigation system hose and sprinkler heads if the fancy "rain" system of the Eurosprinkler is out of your budget range. But, the point here is that you need to be thinking early on about arena sprinkling systems and/or how you will be keeping your arena sand moist as well as leveled.

Okay, so now you have a design for your main horse barn, and you have a design for your arena. Now let's think about why you want to connect them with a **connector aisle between the barn and the arena**. Whether you'll have an arena and whether the barn and the arena are connected are important decisions you'll have to make as early as making a site or plot plan and envisioning whether the plan will fit on the land you are considering. (Remember that I have a connector aisle between the barn and the arena which I discussed earlier.)

I have been told that the cheapest way to construct a barn is to build your horse barn as a shed roof extension structure off of an "A" frame arena roof. It certainly looks like the cheapest way, architecturally speaking. I'll say that much! Theoretically, by building a barn in this way, you eliminate the expense of extensive additional roofing materials, you eliminate one exterior wall and its associated expenses, and you eliminate the need to have a connector aisle between the barn and the arena. For entrance into the arena, you would simply leave out one of the stalls backing against the arena and make that the opening into the arena itself. Of course if you do this, then your horses—or at least half of them—the ones whose stalls back onto the "common" barn/arena wall—will not have windows and will be open to breathing in all the dust that is kicked up by riders and horses in the arena footing and slips in between the wood of the arena side of the stall walls. I have been in barns where they have corrected the problem of windowless, dust filled arena wall stalls by simply not having any stalls that use the common barn/arena wall, but rather, have an aisle that runs along the arena wall with stalls only put along the outside wall. In that way each horse does in fact have a window, or in some cases, a Dutch door, in their stalls and lessened dust issues.

However, by separating your arena and horse barn and using a connector aisle between them, you keep the arena dust, the arena chill factor (the cold winter air from the open arena flowing back into the barn aisles), and the activity separated.

Also, having a connector aisle actually isn't as cost inefficient, in my opinion, as it may seem at first. For example, if you want an arena observation room in your barn, then generally speaking you will need to put the viewing area on one end or the other of the "A" frame barn. That means you will have to build some kind of an aisle extension past the stalls, and then some kind of additional space at the end of the arena for a viewing area. Of course you could use some stall space along the common wall and "open it up" for a viewing area, but I've never seen anyone do that. It seems that invariably, people with "A" frame barns have their viewing area at one end of the "A" or at the other.

Also, in order to have an "A" frame structure that is high enough to allow the roof to slope down over the edges of the arena to form the horse stalls, the side wall of the arena would need to be at least 20 feet high coming off a roof with a minimal 3.5:12 roof pitch. Remember too that the "A" frame would require an extension to the overall length of the arena roof to accommodate a viewing room at one or the other end of the arena.

It would seem to me that by the time you extend and raise the "A" frame roof to make it wide and high enough for stalls, and then extend the overall arena length to make a viewing room, that having a separate barn and arena with a short connector aisle between them, would be similar in materials and labor costs, and far more safe and healthy for the horses, than the singular "A" frame structure. So I'm not so sure the "A" frame arena and stalls should be considered only from a cost saving point of view. Plus, based on the closeness of costs in all the honest bids that I received, regardless of style, it would seem that there isn't much difference in costs between barn and arena styles with similar square footages, but more a difference in costs between various builders. I would suggest that you seriously consider having the barn and arena structures separated but connected by a corridor for the sake of the horses' health. In other words, I would only choose to build an "A" frame if you wanted that kind of barn and arena arrangement for some personal reason.

If you do plan to have a connector aisle, then I would recommend that you plan an exterior, slider entry door at some point along your connector aisle. This will serve you well as both an additional entry option into the barn as well as allowing in additional air flow to the barn and arena during the warm months.

Why would you need an extra slider entry door on the connector aisle when you have end doors on the barn and the arena, and a front entry man door on the barn? Here's yet another example of what can go wrong.

Because the end doors on the barn and arenas are very large and susceptible to blowing in, and potentially off, when there are strong winds, the steel pole barn builders install a centrally located, stabilizer "track" in the ground for the doors. When your doors close, they slip onto that central, metal, holding track as the doors close in the middle. Theoretically, if that track is properly installed in the ground so that the wood post the metal track is attached to is below the frost level, then all is well throughout the winter. However, if the wood post for the track has not been installed below the frost line, then the post can, and will, during a period of deep winter freeze, "heave" up from the ground and jam your barn end doors impenetrably shut! Even if you have a front door to your barn which would theoretically still be functional, if it's not a horse-sized, 4' wide by 7' high door (like the Dutch stall door I used as a front door), but is only a standard, 2' 6" to 3' wide, by 6' 8" high, man-sized door, then you will be challenged to fit your muck wheelbarrow in and out the door or comfortably fit you and your horse in and out the door. But by having a single slider door in your connector corridor that is approximately 6 feet wide by 10 feet high, and one that does not have a centering post on it, you will be able to have access to the outside even if your other big barn doors are jammed shut.

Photo #81: Hay Barn Metal Slider Door Ground Stabilizer (Metal Capped Post)

Our first winter, of course, our center posts heaved and jammed shut all our end doors in the barn and in the arena because, in fact, the posts were not installed below the frost line. We wound up having to buy a power hack saw to cut off the metal slider track caps from the heaved wood posts so that we could open the end doors on the barn. Because we have custom, glass end doors, they are very heavy. The weight of the doors has allowed us to "get away" with not having the stabilizing, metal, center posts on the end doors. On windy, winter days, however, we make sure the center closing chain is snugly attached between the two barn end doors, and the locking hooks on the doors are closed and engaged, which is sufficient to keep those heavy doors from blowing off the barn.

Photo #82: Barn Aisle Glass End Doors

However, if you just have sheet metal for your barn end doors, then the metal of the doors will be too light for you to do without the slider locking center post track. So, if you must have the metal capped post and the post heaves, MAKE the pole barn people return and reinstall the post so that it is properly installed BELOW that frost line! They know better than to have installed the post too shallow.

While we are on the subject of end doors, I would like to make a suggestion for the **end doors on your barn**. Again, where I boarded for too many years, the huge end doors were simply the sheet metal ones that come with your basic pole barn. I can't begin to tell you how often those doors jammed, blew off, blew in, dented, howled, and rattled! They were a constant, dysfunctional annoyance! They were also just solid sheets of metal without any form of light panels in them which made the barn aisles dark when the doors were shut—which is all the time during the winter months in northern climates!

Particularly in the winter, when the horses would come in from the pastures at the boarding barn on a sunny, winter day that had been reflecting sun into the horses' eyes during their turnout from the glistening, white snow, there was a consistent problem with the windowless, sheet metal doors and the dark aisles they created. When the turnout girl would open (if she could) the metal doors and lead the horses into the dark aisle, the horses would ALWAYS panic—usually only briefly—but definitely panic. What made the panicking worse was that the horses' hooves were invariably jammed up with ice balls from the pasture snow, which in turn made their feet slippery, which in turn greatly increased the risk of injury when they panicked upon entering the barn.

To avoid these two issues—dark aisles and panicking horses—I was unswayable in my determination to have end doors with glass windows. And I'll advise you right now if your barn will be enclosed with end doors be sure to put custom, glass, end doors into your budget, regardless—you will LOVE them! Not only do they give pleasant light to the barn, and eliminate the transition panic for the horses from the outside to the inside, they allow everyone—horses and people—a beautiful additional view to the outside world. The other "hidden" additional and important benefit—already mentioned—is that the heaviness of the custom glass, barn end doors makes them more stable in the wind, and therefore significantly more functional over all!

By the way, I finally found information as to why it was that the horses panicked so badly when transitioning from the bright snow into the darkened barn. We humans know that this happens even to us if we've been out in the bright snow and then suddenly come into a dark area. Such a transition makes us go temporarily blind. Well, that's the same phenomenon that happens to the horses. However, because the horse's eyes are so much bigger than our eyes, horses are more severely affected by such transition in lighting and it takes much longer for the horse eye to stabilize after such a dramatic light change than it does our eyes. Therefore, the horse is literally blind for much longer than we would be in such a transitional lighting situation. When the slow light transitioning eye of the horse is combined with the horse being a prey animal, the combination causes horses to flat out panic! This is also why proper uniformly illuminated arena lighting is so important which I will discuss later.

So now hopefully, through all this reading, you've come up with your own "dream" barn and arena plan on paper and even have a feel for how it will all look three dimensionally.

Pretty exciting, isn't it?

You also have a LOT of information about things to consider having in your barns, as well as information about the many pitfalls to avoid.

As "finished" as your mind might feel it is about planning anything else, incredibly, there are still lots more things to talk about and think about before construction actually begins so that you can add these items upfront and cost effectively. Some of these considerations are optional, but some are essential.

Chapter Eight

Finishing the Horse Barn Interior

Let's first talk about the general essentials that need to be considered. Pull out that barn plan yet again, as well as your survey and site or plot plan on which you've marked your building placements, driveways, and pastures.

Now let's talk about **general plumbing**.

First, you should realize that your water feed lines run from your well to your house and then to your barn and then from either the house or the barn to other places where water is needed. The house is generally where the pressurized tank is located that stores water from your well unless a larger external storage tank is required in locations that have low flow rates from the well. There is a split on the pressurized water storage tank in your house that either allows water to flow into a water softener or to bypass the water softener. The point is that not all plumbing fixtures and uses of water need softened water. The barn, like the house, however, does need both softened and unsoftened water. Therefore, both types of water lines need to be run to the barn from the house. These water lines need to be laid in a trench that runs from the house to a utility room in the barn. Once inside the barn utility room, the softened line feeds into your barn's hot water heater plus the plumbing fixtures where you want softened water. Specifically, in the barn, softened water would be in the wash rack, the bathroom (sink and toilet), utility tub, optional clothes washer, hot water heater, kitchen sink (but not the refrigerator if it has an automatic ice maker).

The **un**softened line is used to feed water to the automatic waterers in the horse stalls and those out in the pastures. You don't want the horses drinking the salts that make the softened water soft. **Un**softened water is also fed into all freeze proof hydrants inside and outside the barn and to your automatic ice maker in your refrigerator if you have one.

Next you will need plumbing for **hot water heating in the barn floors** if you choose to have heat and choose this safe form of heating which I strongly recommend you choose if you decide to heat your barn. In this system, the hot water heating tubes run inside the concrete floors in the barn in such rooms as the office/viewing area, the bathroom, the tack room, and potentially under the barn aisle floors if you decide to heat the barn itself outside of the heated rooms. This tubing needs to be laid down on 2 inch thick

Styrofoam boards over 4 mil visqueen sheets **before** concrete is poured for the floor and over the tubing.

NOTE: The plumber runs all the water lines and plumbing lines in early "rough" construction and then returns near the end of construction to install the "finish" plumbing that includes fixtures like sinks and toilets. However, you should select and order your plumbing fixtures early in the construction because the delivery of the fixtures is slow.

NOTE: As previously mentioned, I strongly recommend a free standing sink in the barn bathroom such as a pedestal sink without a cabinet underneath to minimize mouse hiding places.

Photo #83: Barn and Enclosed Rooms In-Floor Heating Tubes before Aisle Cement is Poured

Earlier in the project planning which included doing the barn floor plan, you had to decide on the location of your wash stall, which should be located either next to the heated bathroom, the heated tack room or the heated office area so that the hot and cold water faucets in the wash stall come out of a heated wall into the wash stall. This will keep the hot water as warm as possible in its run to the wash stall and cut down on the expense of heating the water as well as preventing the water lines from freezing.

If you have decided to have **automatic waterers** in your barn, you need to plan ahead for the water plumbing lines to run into each stall. If you also want automatic waterers in your pastures, you will have to determine where you want those located so that plumbing lines can also be run out to them. All these water lines both inside and outside the barn must run below the frost line (which is 4 feet below the ground surface where we live). Again, you must make sure that you have **un**softened water lines feeding the automatic waterers.

Inside the barn, you will need to consider if you want to share plumbing lines between the stalls for the automatic waterers (which means you would have one plumbing line run from the barn utility room to serve each pair of two automatic waterers with one waterer back to back on either side of a shared stall wall between the stalls) or individual lines run to each waterer. I ran individual lines to each waterer, and do not have the waterers placed back to back on the stall separator walls. I felt that if a plumbing line broke or froze or a waterer needed to be replaced, that it would be better to have just one of the waterers shut down until a repair could be made, rather than to have all waterers shut down. Naturally, it is cheaper to share the plumbing lines than to run individual ones.

UPDATE: Well, now we have found yet another reason why you should seriously consider NOT placing the automatic waterers back to back on stall separator walls. The other morning we came into the barn to find that the counterweight in the automatic waterer in Ian's stall did not rock back into place and stop the flow of water into the bowl. As such we had the soggy stall and flowing river problem occur once again. But the point here is that this time, when the waterer malfunctioned, Ian decided to take control of the situation. He used his "horse brains" to solve the problem of the misbehaving waterer and just kicked the crap out of the waterer … as well as the stall separator wall between his stall and Emma's stall next door. Not only was Ian's waterer damaged and dysfunctional, but the nice, thick, pine boards along the top of the waterer were split open and daylight was coming in thru the split from Emma's stall. If Emma's waterer had been mounted back to back on the stall separator wall, then her waterer would most likely have been damaged and dysfunctional as well as Ian's. I should also take a minute to note here that this is yet another reason for having that "one extra stall" available!

You will also need to decide where you want the waterer to be located in the stall. Do you want to have it in a back corner? Do you want to have it in a front corner? Do you want it along the stall wall? If you want it "along" the stall wall, then exactly where on the stall wall do you want it? Halfway down the stall wall or one third of the way or perhaps one quarter of the way?

I placed my waterers approximately half way down the stall separator walls and I am happy with that. I felt it would be too cold to have the waterer in the back corner of the stall by the outside wall where more heat would be required to keep the bowl water heated to the appropriate, palatable water temperature. If the waterers were in the front corner of the stalls, I felt the horses would feel uncomfortable about having to come into full view every time they wanted a drink. I should mention that I have the **feeding bowls** mounted in the front corner of the stall on the same wall as the waterers. The location of the waterers and the feeding bowls has worked out perfectly for the horses. It seems to have the correct logic flow to them for how they access their food and water sources.

I should also mention here that I paid extra money to buy heavy duty, stainless steel feeding bowls that match my automatic waterers. I really love the nice, big, solid, and essentially indestructible feeding bowls. They are also designed for easy removal for regular cleaning as are the water bowls. I strongly recommend you pay the extra money for these heavy duty bowls and NOT buy plastic feeding bowls. I also have **salt blocks** mounted on the same wall as the automatic waterers and feed bowls, 2½ feet along the wall from the stall front. I made the mistake of getting vinyl coated salt block holders—they only lasted one year before being miserably beat up. Look for stainless steel salt block holders. I have been told that the automatic waterer manufacturers are developing a new salt block holder that just might be worth waiting for!

NOTE: Automatic waterer bowls come in different sizes. Be sure to get a big sized bowl if you have big sized horses.

Photo #84: Automatic Stall Waterer and Matching Stainless Steel Feed Bowl

For review: Facing the stall front, the feed and water bowls plus the salt block are all on the left side of the separator stall wall as is the cut out in the stall front for feeding. The feed bowl is a corner mounted bowl. The salt block is 2.5 feet from the stall front (although not shown in **Photo #84**). The waterer is 5 feet from the stall front to the center point of the water bowl. The top of the feed bowl is 3′10″ above the stall floor which puts the bowl right under the feed slot cut into the bar grill on the front of the stall. The top edge of the salt block holder is 4 feet above the stall floor. The top of the automatic waterer bowl is 3 feet above the stall floor.

NOTE: Horses must have free access to a salt and mineral, salt block. The salt and minerals are essential to the horse for overall good health.

Determining where you should place your waterers in the pasture has some important considerations. First, you should have the shortest possible distance for the water line run from the barn, out to the pasture waterers, for cost effectiveness. It is also important to be aware that the waterer should be placed far enough off the pasture fence so that at least two horses can easily gallop between the fence and the waterer. I placed my waterers 22 feet off the fencing. It is also a good idea to place the waterer a significant distance away from the pasture entry gate. Horses are often highly animated, and sometimes even panicky by the gate such that having the waterer within kicking or trapping distance near the gate would be a mistake. Nevertheless, you don't want to put your waterer too far away from the in gate either! This is another mistake I made in one of the pastures. I made it a hideously long walk to one of the waterers which makes daily clean out a real chore! Placing the waterer about 68 feet from the entry gate into the pasture works well.

NOTE: Waterers should be cleaned out every day. It's very simple to remove and replace the bowls as well as easy to wipe them clean. Whereas you can skip a day in cleaning inside water bowls, you cannot skip a day with the outside bowls. The outside bowls fall victim on a daily basis as a minimum to all kinds of drowned bugs and bird droppings and therefore must have a daily clean out!

However, you should also consider whether or not you really need or want to have automatic waterers in your pastures. If you aren't going to have your horses in the pastures all day and/or all night, then you may not really need to have automatic waterers. Remember that outdoor waterers need daily cleaning. Also, the heaters run up the electric bill in the winter months when the heater is fighting to keep the water temperature palatable against the outside, subzero temperatures. However, if your horses are going to be out all day and/or all night, then these waterers are wonderful. By having the automatic waterers in the pastures, you don't have to worry about your horses having a constant, temperature appropriate source of water while turned out.

NOTE: The automatic waterers require plumbers AND electricians. Remember that the waterers have thermostatically controlled heating elements in them to keep the water from freezing. These keep the water temperature at about 55 degrees. Therefore, electrical lines will need to be run to each waterer (inside and outside) and placed in the same trench dug to run water lines to the automatic waterers.

NOTE: If you aren't considering automatic waterers in your pastures but will be turning out your horses all day and/or night in the winter, then you will still need to have trenches dug for underground electrical wires to be run out to each pasture to mount at least one double electrical outlet on at least one post of each pasture. These outlets will be used to plug in heated water buckets or for heaters that are placed in water troughs for the winter months when needed.

Although I discussed the plumbing needs for aisle heating in your barn, I would now like to offer you some information to help you consider whether or not you even want to **heat your barn aisles**.

So, let's take a minute here to consider reasons why you should, and why you should not heat your barn.

NOTE: You should of course heat specific areas in the barn as I mentioned earlier including at least your office/lounge/viewing area, your bathroom, and your tack room. Also, I strongly recommend you use the in floor, hot water heating system for these areas regardless of whether or not you decide to heat the barn aisles.

As you know, we heated our barn by using the hot water, in floor, tube system fired by a gas hot water heater and heat exchanger. This is the safest possible heating system you can use in a barn as far as a fire factor is concerned. When we started our planning and construction, Hurricane Katrina had not yet happened. We also hadn't realized that getting a public utility natural gas line run into our farm would be basically impractical. So we thought that by having a propane gas fired, hot water based, heating system in the barn, it would be very economical to heat the barn. By "heat the barn," I don't mean that we were looking to actually "heat" the barn. We just wanted to be able to take off the chill in the barn on those single digit, and below zero, winter days and nights. However, things didn't turn out the way we hoped.

Because of the need to have major air flow in your barn—as we discussed earlier—the concept of heating such a structure is essentially an oxymoron. In addition, I have found that even with only three horses, and the good air flow, high ceilings, and the open air exchange between the arena and the barn, that the temperature in the barn tends to stabilize at between 16 and 18 degrees even on the minus 5 to minus 10 days we have had, and whether or not I had the aisle heat turned on or off. The temperature stabilization probably has as much to do with heat escape from the heated rooms in the barn as much as from any body heat generated by the horses. However, I would imagine that without heat anywhere in the barn, and therefore having no subsequent heat escape, that such a barn would get significantly colder than ours has. In the barn where I boarded for years, even though they had neither aisle heat nor significant heat escape from heated rooms, the barn temperature also consistently stabilized like ours has at about 16 to 18 degrees regardless of the outside weather. However, in that situation there were about 40 horses on the aisle, each of whom was pumping out a lot of body heat!

The bottom line biggest problem with trying to heat the barn aisles is the phenomenal cost of the natural gas or liquid propane gas required to heat the barn to any degree. Our gas bill was double what we expected—about $8000 per year and prices are still going up. Remember too, when considering heating the barn, that when it gets really cold, there's always that quick and easy solution available to keep the horses warm called putting on an extra blanket. In fact, after the first winter's heating bill, I purchased medium weight, turn out rugs that I add as a third layer if I think it's going to be brutally cold in the barn, and it's worked out just fine now that I avoid turning on the aisle heat. So, considering the temperature stabilization from heat escape, and the gas bill, and the availability of horse blankets, whether or not to heat the barn aisles is once again, your call.

However, if you have low ceilings and can close off the barn aisles with doors, and want heat in your barn, it is possible.

<u>NOTE:</u> However, you must not forget that you HAVE to have big air flow through your barn or you will have a urea vapor nightmare and unhealthy air for the horses. So no matter how you try and heat the barn aisles, it will never really be energy efficient or cost effective.

The system I have is a good one again because it is as safe as heating systems get for barns regarding not being a fire hazard, it is relatively cost efficient, and it provides uniform heat. Heating systems such as forced hot air systems have an inborn fire hazard in barns due to the huge amount of barn dust that would come in through the intake returns. To be more specific about my system, it is called an in floor, hot water heating system. The heating tubes run under all the cement floors of the aisle, but do not run under the stalls because of the stall base and mat system I chose to use. If you decide to have cemented stall floors, then the heating tubes could also be run under the stall floors. In my barn, the tubes also run under the cement aisle to heat the tack room, bathroom, and the office/viewing room areas.

Preparation for the in floor heating system requires:

(1) laying 4 mil sheets of visqueen down on the substrate soil;

(2) laying 2" thick styrafoam blueboards (4' x 8') over the visqueen;

(3) laying the tubing for the "hot water" lines;

(4) pouring cement over steps #1, #2, and #3.

There is a heat exchanger/hot water boiler that is placed in your utility room. The heating tubes are actually filled with antifreeze (not water). The hot water from the gas fired hot water tank flows into a heat exchanger that heats the antifreeze flowing through the heat exchanger. The antifreeze is pumped through the tubes in a constant state of recycling, passing through the heat exchanger and then into the tubing in the floor. In the utility room where the heat exchanger is located, you will also have your control panel. The control panel lets you turn different zones on or off for heating as you desire and have predetermined.

I have three heating zones in my barn so that the thermostats in each zone can be set at different temperatures or turned off. There is one zone for the tack room, one for the barn aisles, and one for the bathroom and office/viewing room areas. In order for the heating tubes to cross the aisle for the zone that heats the tack room, the plastic water tubes had to run across the top of the connector corridor aisle walls. A wooden "bridge" was built at the top of the corridor walls to hold the water tubes, and then foam insulation was sprayed over and around the tubes. After the insulation was sprayed on the tubes, our builder was supposed to finish the bridge by boxing in the sides and top of the bridge to cover the tubes and foam. However, that didn't happen. Although I am disappointed that the bridge doesn't have a finished look to it, the good side of the unfinished nature of the bridge, is that the cats get to use it as an aerial crossover bridge when they are mousing from above! **(see Photo #75)**

Photo #85: Spray Foam Insulation

Photo #86: Spray Foam Insulation on Exterior Wall of Barn Office

NOTE: The same spray in foam insulation used to cover the heating tubes that run across the aisle on the bridge, was used to insulate the walls and the ceilings of our office/lounge/viewing and utility area, the bathroom, and the tack room. As I previously mentioned, spray in foam insulation is a thorough, easy way to make an air tight insulating seal for heated rooms. The low density foam is only slightly more expensive than fiberglass batt insulation, but is much more effective in its insulating ability.

I've been in show barns that are heated with gas fired, hot air space heaters that are suspended from the ceiling. This type of heater does indeed pump out a lot of directed heat. However, the problem with them, as with anything and everything that stands, sits or hangs in a barn, is DUST! Because the accumulation of dust is so extreme in a barn, if you don't have time to regularly clean every surface of this type of heater, they are, in my opinion, a fire hazard with a fire just waiting to happen.

There are also geothermal systems as an alternative to the type of heating system that I installed. A geothermal system might be a more cost effective way to heat a barn these days since gas prices have soared. However, I don't have any direct experience with this type of system, but I do think it might be worth your while to check it out if you want heat in your barn.

Photo #87: Spray Foam Insulation on Aisle Wall of Barn Office

Now, let's get back to the plumbing.

You will need to have freeze proof water hydrants both inside your barn and outside. Once again your plumber will be installing these and running the water lines to the pumps. These pumps are needed for many reasons, including filling buckets should the waterers fail, running hoses into the arena for watering the footing, putting out fires, washing off equipment, etc. You need to determine where you want these hydrants placed both inside and outside your barn.

Inside the barn, the placement and number of freeze proof water hydrants will be relative to the length of your aisles. If you have a relatively short aisle as I do (72 feet), then one, centrally placed hydrant is all that is necessary. If you have long aisles with lots of stalls, you will want to have several hydrants placed along the aisles and spaced so that your hose at no point will have to stretch more than 50 feet. In this way, you will have a manageable length of hose with which to work and that will also keep the filling of the water buckets more manageable.

Be sure to place the hydrants along the aisle so that they do not interfere with putting grain into the feeding bowls or with opening the stall doors or with handling the horses in the aisles--or interfere with any other tasks you will be doing. You will also need to place a hydrant that enables you to reach your arena with a hose from the hydrant. So, if your connector corridor is long, then you might need to place a hydrant at the end of the connector aisle near the arena, but placed so that it doesn't interfere with the entry gate or placed where a panicked horse could get caught up on it.

Outside, you will need to have a hydrant near your hay barn in case of fire. You will also need to have freeze proof hydrants to provide access to water on either side of the main barn, near the pastures. You will also want an outside hydrant that can be used to wash off your farm equipment. In my case, I placed my north side, exterior hydrant along the turnaround driveway in front of the hay barn so that one hydrant serves multiple purposes. Then I have another hydrant on the other side of the barn next to the south pasture for water access at that end of the farm.

If you are going to have an outside wash rack, then you will need to have a freeze proof hydrant located next to the wash rack.

In your office/lounge/viewing area, you will basically need to determine which of the following you will be including in the room:

- If you are going to have a kitchen area, you will need plumbing for the kitchen sink faucets and for the refrigerator if it has an ice maker;
- If you are going to have a washer and drier, you will need plumbing for the washing machine;
- If you want a dedicated tack cleaning area, you will need a utility sink.

At one point I considered putting a sink in the tack room for cleaning tack. However, I'm glad that I didn't do that. Nevertheless, if you want one, be sure to plan in the water lines for hot and cold water into the room and a utility sink.

One last note regarding the plumber is that the plumber installs the septic pumps in the septic system. The electricians will then hook up the electric power that runs the septic pumps.

As overwhelming as planning your plumbing may seem, there is an even more pervasive utility to now consider–**electricity**. All the electrical items from wiring to switches to installation of lighting fixtures will be handled by an electrician, but the capacity (number of amps) of service, the type of fixtures and outlets, and all placements of electrical items will be determined by you.

As we discussed previously, you will need to have 400 amps of electrical service on your farm which the electric company will run into your farm into a transformer that they will install on your property, and then install the wiring that will connect the transformer to the service entrance (the place where the electrical service enters your house) with a meter on your house. Then your electrical subcontractor will complete the wiring inside your house that runs from your house to the barn and other buildings. You will also need to have the excavator dig a trench that runs between the house and the barn/outbuildings in which to lay the electrical service wires (as well as all the other service pipes and wires).

For our farm, the 400 amps of electrical power fed into our house is split into two breaker panels in which each breaker panel has 200 amps of service. One of the 200 amp breaker panels not only supplies 100 amps of electrical service to the barn, but is also a "standby" breaker panel. The remaining 100 amps that do not go the barn on that 200 amp standby breaker panel are used to supply electricity to the critical utilities in the house such as essential, minimal lighting, the well pump, the water softener, televisions, the refrigerator, garage door openers and the furnace fan. This 200 amp standby breaker panel can also be connected to a liquid propane (or natural gas) fired **generator of about 12 kilowatts of capacity** for automatic switch over to back up critical power needs during power failures. The other 200 amp breaker panel operates non-critical utilities in the house such as the air conditioner, non-critical lighting, and the other kitchen appliances such as electric ovens.

The 100 amp service to the barn supplies a 100 amp breaker panel in the utility room of the barn. All the horse barn, arena, and hay barn power comes from this panel (which is all potential standby power if you have a switch-over generator).

In order to run electrical power to the pasture waterers, the hay barn, and the driveway column lights (or other sites where you will require electrical power) from the breaker panel in the horse barn, trenches must be dug from the horse barn to these various sites in which wires will be laid for electrical service to these sites.

NOTE: The main electrical load on the barn's breaker panel is the arena lights which must be split into several circuits; the number of circuits depends on how many lights you have in the arena. I have 28 arena lights of 400 watts each and 4 separate circuits. What is important to remember is that when your system goes into standby mode due to a power failure, you must **not** have on your arena lighting circuits or they could potentially overload your standby generator and cause a system failure as well as potentially cause a fire. Only one circuit of arena lights should be turned on during a power failure when your electrical service is being run by your standby generator.

I think the simplest way to discuss electrical needs beyond the aforementioned service basics will be to start envisioning your needs at the floor level of each structure and then envisioning "up" from there until we reach the ceiling. Let's start in the barn and then talk about the arena and closed rooms separately. Lastly, we'll talk about exterior lighting.

NOTE: All your electrical wires lower than 8 feet off the ground in your barn and arena must be encased in metal conduit tubing (EMT = electrical metal tubing) so that horses cannot chew on the wires and get injured. The wires also need to be protected in EMT from accidental damage that may occur from such things as farm equipment in the aisles. Wires not encased in EMT are potential sources of electrical injury as well as fire hazards.

Basically you will need to have an outlet box installed on each and every horse stall, as well as the wash stall and the grooming stall. The outlet boxes should be located on the aisle side of the stall separator walls. These power sources will be essential in summer to power the box fans on the stall fronts. They will also be used for myriad other applications from clippers, to plug in air fresheners, to vacuum cleaners! You really can't have too many outlet boxes and should plan them in everywhere.

Photo #88: Water Consumption Meter and Stall Electrical Outlet

Next you will hopefully be installing the water consumption meters for your automatic waterers also on the aisle ends of the stall separator walls which will require the electrician to run a wire from the waterer to the consumption meter. The electrician will also be the one who wires up the heaters in the automatic waterers. Of course, the wiring for these heaters was required in the trenches run from the barn utiltity room to each stall that also contained water lines. As noted earlier, after our automatic waterer flood, I now would suggest having warning bells in your house for times when the waterer fails (run through telecommunication pipe running between the house and the barn), and a water sensor in the floor under the waterer with an automatic shut off switch (which the plumber can install). It is also very important to insist upon having one dedicated circuit per automatic waterer, both for the inside and the outside waterers. This will allow you to shut off each individual waterer should one need repair, should certain stalls not be in use or should you not be using the outside waterers in the winter or only certain ones.

You will also need to have power outlets in the tack room, the bathroom, and the office/lounge area. In the office/lounge area you will also potentially need a heavy duty 240 volt power outlet for your air conditioner hookup (if it is a large size air conditioner—and be sure to give your builder a "heads up" for where the air conditioner hole should be cut in the wall!), and potentially your clothes drier hookup (if you're using an electric drier and not gas), and 120 volt outlets for any full sized kitchen appliances you might have including refrigerators, dishwashers, garbage disposals and the like.

Of course, each area will need light switches. Placement of light switches should not be taken lightly or left up to the electrician's judgment. You will want to thoroughly think about exactly where you want your light switch boxes to be located and where you want redundancy. As an example, ask yourself if you only want to be able to turn on the barn aisle lights from the front door of the barn. Or if you also want to be able to turn them on or off from within the office area. Or if you also want to be able to turn the lights on or off from either end door of the barn or at the end of the connector aisle by the arena. How about when you enter your office? Do you want to be able to turn the lights on and off from either entry door? And what about your exterior lights? Where will you want switches to turn exterior lighting off or on? This is the kind of thinking that you should do when planning where your switch boxes will be located and where you want redundancy.

You will also want to make sure you have enough switches available on the switch plate for lighting certain areas. As an example, let's discuss my grain bin area which I have in the foyer of our barn. Here there are four switches. One switch turns on the outside porch light. One switch turns on the chandelier in the foyer. One switch turns on the barn aisle lights. Then, one additional, very important switch turns on halogen spotlights that direct light down right into the opened grain bins so that I can actually SEE what is in the bins as well as what I am doing while scooping out grain and additives! **(see Photo #54)**

So not only do you need to think about where to put the switch plates, redundancy of switches, number of switches on each plate, but also where you want/need special, focused lighting.

Now let's visually move up the walls to think about the lighting needed on walls and ceilings and the corresponding lighting fixtures.

Nothing was more frustrating to me when horse showing than coming into a show barn in the early morning hours to braid my horse, to find myself in such a dark barn with even darker stalls, that I could barely find my horse in the stall no less try and braid him! I eventually resorted to wearing a miner-like head lamp and hanging battery operated camper lanterns on the stall walls to try and see what the heck I was doing with the horse's mane! Of course none of this portable lighting was particularly convenient or non-spooky for the horse!

Then too there was the day when I had finished braiding and grooming my "brand new" gray horse for his first show, tacked him up and brought him out of the dark, dank show barn to my anxiously waiting trainer for her to do the warm up class for me. My proudly beaming smile was instantly wiped off my face when the trainer snarled at me and said with breathtaking haughtiness and anger, "Don't you ever bring me a horse with 'green' on it again!" Well, in the first place, I had never had a gray horse before, and so I had no idea what in the world the trainer was talking about when she snarled out the "green" word. But later, I came to find out that "green" is trainer code for horse shit. But, case in point here is to let you know just how frickin' dark the show barn was. Even though I had thoroughly groomed the horse after braiding him and before tacking him up, I never saw the huge patch of horse shit stain on his rump! Oh, well. He won a blue ribbon anyway, poop stain and all!

So, based on such "dark" experiences, I vowed that when I built our barn, that there would be plenty of light—day and night—in both the aisles of the barn and in each individual stall so that a missed patch of "green" would never get by me again! Also, having good lighting available day and night allows you and your vet to be able to take care of your horse should he or she become injured or ill. You can also muck out the stalls any time of day or night and be able to see what you are doing. So aisle and stall lighting are very important.

Above each stall I have a singular, instant on, to minus 20 degrees, 11 and 3/8 inch diameter, round, enclosed, water tight gasketed, clear polycarbonate lens, fluorescent **"stall" light** fixture with a 36 watt fluorescent lamp and an electronic ballast in each that I purchased from an equestrian barn supplier made specifically for stalls to economically distribute a proper amount of light. There is one stall fixture mounted on the ceiling above each stall. The fixture must be mounted on the ceiling above the stall (not on the stall wall—which of course is where our electricians first merrily mounted our stall lights) in order for it to properly throw the light. **(see Photos #39 and #65)**

The only fixture most barns have for lighting the barn is a single incandescent light bulb, enclosed in some kind of protective wire cage, mounted on the top of a stall front along the aisle, with only one light mounted for about every 3 or 4 stalls, and no lighting at all in the stalls. Zoicks. No wonder I was wearing head lamps to braid!

NOTE: My roof pitch and exterior wall height combine to put my stall light fixtures approximately 12 feet above the stall. If you will be mounting your lights higher than that, I would suggest considering getting two stall lights per stall.

I do not have individual light switches for each stall light although that would have been nice. What I do have is a switch on the aisle end of the first stall front that lights all the stall lights in that line of stalls on that side of the aisle. On the other side of the aisle I have an equivalent light switch for the ceiling stall lights for the 3 "future" stalls.

Over the wash stall and the grooming stall where I wanted to have more light, I installed two of these circular stall light fixtures per stall. Each of these stalls has their own light switch on the aisle end of the grooming stall and the wash stall, respectively, so they can be turned on independently. I had wanted to have four lights per grooming and wash stall, however, but I got voted down on that request. What I have learned is that there is plenty of light having just the two fixtures in the wash stall where the light readily reflects off the bright, white walls. However, in the grooming stall, where the walls are lined with pine, the light is not nearly as bright as I would like it to be. I would like to suggest that you avoid my mistake and have four lights installed for the grooming stall—two lights placed to light the back of the stall and two lights placed to throw light in the front of the stall.

For the foyer, I had a little fun and put in a **chandelier**. I think putting a chandelier in your barn makes a statement of just how important your horses, their housing, and your riding are to you. It tells the world that they aren't just in any barn, but they are in your barn where you care about the barn and everything in it--especially your horses. Of course, selecting a "barn appropriate" chandelier is a good idea. In other words, the chandelier should be something that is very laid back, rustic looking, and one that is able to take a lot of dust without showing the accumulation too obviously between dusting sessions. I chose a colonial, wrought iron fixture with a kind of basic "blacksmith" look to it. **(see Photo #54)**

For the aisle, I suspended 4′ x 7″, 3 lamp, fluorescent fixtures, enclosed and water tight gasketed, low temperature start electronic ballasts, providing 84 watts of fluorescent lighting per fixture. These lights hang down from the highest point of the barn aisle ceiling from cables of a length to put the light fixture at about 14 feet above the aisle floor. There is one of these lights spaced every 12 feet on both the barn aisle and the connector aisle.

Photo #89: Suspended Barn Aisle Fluorescent Lighting, Barn Aisle Eave Ceiling, Box Fans on Stalls

In the tack room there are two, ceiling mounted, enclosed, 4 lamp, fluorescent fixtures. For the bathroom I just used two down lights above the sink which brightly light the entire 5′ x 5′ room. In the office/lounge/viewing area, there are three lighting zones. There are two down lights over the utility sink/washer/drier area that are on one light switch. There are two enclosed, ceiling mounted, 4 lamp, fluorescent fixtures on the ceiling in the main office/lounge area on their own circuit and switch. There is one enclosed, 4 lamp, fluorescent fixture in the viewing area on its own circuit and switch. The office lights can be turned on by either entry door into the office. The entry/exit door to the exterior of the barn from the office also has a switch included on the switch panel for the light fixture on the outside of the exterior wall by the entry door. The light switches by the entry door into the barn aisle from the office has one switch for the ceiling light fixtures in the office area, one switch for the fixture in the viewing area, and one switch that turns on all the barn aisle and connector corridor lights. The switch for the down lights in the office over the washer/drier/utility sink area is located on the wall by the man door into the utility room.

Upon entering the utility room, there is a light switch that lights the single, incandescent light fixture on the ceiling. Inside the utility room are the electrical service panel, the hot water heater, and the in floor heating system plumbing with its control system pumps, valves, and heat exchanger.

Also very necessary are exterior electrical outlet boxes. For those of you who celebrate Christmas and like to decorate with lights, you might like to have specific Christmas decoration outlet boxes. So plan now on the outside of your barn where you will need outlet boxes to light up Christmas trees and wreaths and the like. There are also a lot of electric power tools you will be using that will need exterior outlets. Be sure to plan the location for these as well. I have one exterior outlet box located next to the front entry door to the barn, one by the office porch entry door, and one on the outside of the hay barn. Although I have a sufficient number of exterior electrical outlet boxes, putting in extra electrical outlet boxes is always an "extra good" idea!

NOTE: If you want to do any kind of dramatic wreath display for Christmas on the side of your barn above your end doors, you could have your electrician install an electrical outlet placed on the siding of your barn right up where you will be hanging the wreath so as to facilitate plugging in the wreath without having to have 50 feet of accompanying power cord running up the side of your barn to light the wreath!

You will also need decorative, exterior lighting fixtures by your entry doors into your barns. Once again, you will want to select something barn friendly. Again, something either colonial or rustic looking tends to work well.

NOTE: Don't forget to order a coordinating exterior light fixture for the entry door into the hay barn!

Before we begin discussing arena lighting, there are a few more important lighting concepts and electrical necessities to consider.

Photo #90: Hot Wire Junction Box on Fence near Hay Barn

First, if you have chosen to have flex fence for your pasture fencing, then you will need to have a hot wire running along the top of the fence to keep the horses from leaning and rubbing on the fencing which would stretch the plastic tape. That hot wire needs to connect electrically to either the main barn or the hay barn. Plan your junction box location for your hot wire on your fencing close to either the horse barn or hay barn. The hot wire (high voltage) junction box will require having an underground wire run to it from the main electrical panel in the horse barn or from a circuit in the hay barn. You can always just plug the hot wire junction box into an exterior electrical outlet box by using an above ground power cord, but having a wire running along the top of the ground will carry a tripping and entangling hazard with it.

Photo #91: Driveway Column

Second, you should think about your entrance into the farm. If you have decided to have full perimeter fencing, then you will need some kind of a gate for entry into the farm to complete the full enclosure of the full perimeter fencing. You'll remember that I do have full perimeter fencing, and although I haven't as yet found an affordable entry gate that looks "right" for my desired effect, I do have my brick columns installed and have temporarily suspended a utilitarian farm gate between the columns. The point here is that on top of each of the entry gate columns is a light fixture. In order to be able to have lighting fixtures on top of your columns, you must have planned ahead and run electrical wires trenched out to where the columns will be built. You should also plan for the lights on top of the columns to have **light sensors** on them so that they can automatically turn on at dusk and off at dawn which means you will need a junction box on one of the columns where the light sensor can be located. The column lights should also have a dedicated switch in the barn so that you can turn off the column lights any time you might want or need to do so.

Photo #92: Driveway Columns with Gate

In addition, you will probably also want to have electrical lines trenched out to the columns to the junction box for an electric gate opener. Having to get out of the car to open the gate, and then having to get back into the car to drive through the gate, to then get back out of the car to close the gate, to then have to get back into the car to drive to your garage, gets to be a little cumbersome over time. I think it's rather obvious that being able to open your gate with just a touch of a clicker would be highly desirable. If you have an automatic gate, you will then also need some kind of speaker and/or video system with a keypad and/or remotely controlled opener so that you can open your gate from either the barn or house when someone other than your family needs access to your farm. There are so many options for your automatic gate accessories, including everything from hard wired to wireless systems, that it is best you discuss the options with suppliers to determine which system you would like to use, and then determine what power sources you will need to have, and where you will need to have them, to accommodate the option you select.

Arena Lighting is important to do correctly. Remember how we discussed the horses' eyes being slow to react to changes in light level when they entered into the darkened barn aisle from the bright, sunlit, snow covered fields in that facility where I boarded? Well, there was another consistent problem at that facility that I could never quite figure out the cause of until we researched proper lighting for a riding arena.

At the old riding facility, every time I would have a night riding lesson with the arena lights turned on, there were consistent problems. No matter what horse I was on, there were places in that artificially lighted arena that caused the horse to either freeze up, scoot sideways, outright spook, or refuse to jump a jump. Of course the trainer was always quick to blame me for not having enough attitude, blah, blah, blah … However, I knew that wasn't true because I carried the same attitude no matter where I rode or when I rode, and that same attitude at shows won me lots of blue ribbons. Plus everyone else had trouble with their horses in the lighted arena in exactly the same spots as me. So, I knew there was something more to the horses' behavior in those "certain spots" than my attitude.

Once we had researched the arena lighting for our barn and found out how important the need for evenly distributed, bright lighting was in an arena for horses, I went back to the boarding facility and took a long hard look at the lighting they had in their arena. Yipes. First of all, even when all the lights were on (which was rare because the owner wanted to save money), there were clearly several "dark spots" in the arena. And guess where those dark spots were. You're right! They were exactly in those places where my horse would always spook (and other people's horses as well).

In addition, there were "hot spots" (brightly lit areas) due to the totally random placement of the lighting, as well as the variation in the type of lighting fixtures and their wattage on the arena ceiling.

Okay, so there in the arena in that old barn were brightly lit spots and dark spots as you went riding along. When you're going to a jump, you and your horse are traveling forward at a speed of somewhere between 14 to 18 miles per hour. Now add in the eye of your horse which is slow to transition between varying light levels. What do you think was going to happen in that randomly lit arena at the boarding barn? Think maybe your horse might freeze up, scoot sideways, outright spook, or refuse to jump a jump? That'd be right!

So here's the real deal on how your arena lighting is supposed to be. First of all, again, you are trying to achieve equally bright lighting across the entire arena without dark spots or shadows.

Calculating the correct amount of lighting is based on the geometry of your arena—in other words, proper arena lighting is achieved by multiplying the length of the arena times the width times the height (l x w x h)—and the type of lighting fixture that you select. Armed with your arena dimensions and the lighting fixture you want, go to the website **http://www.stoncolighting.com** and put your arena dimensional data into their lighting calculator which will tell you the number of fixtures you need and their placement on the arena ceiling in order to achieve the number of foot candles of lighting you want to have. Or you can get a feel for the calculator by using our lighting selection and arena dimensions (72′ x 136′) by navigating on the website to industrial lighting, low bay 22″ multi-bay series, and select the file Mbl400ma.ies in the QuickEST calculator. In general, the amount of lighting for an arena is:

Minimum lighting	=	10 foot candles
Average lighting	=	20 foot candles
Small show lighting	=	30 foot candles
Best overall lighting	=	50 foot candles

In my 72 x 136 foot arena, I have 28 light fixtures for a total of 76 foot candles. I used 400 watt, metal halide, 22" low bay reflectors for the lights. Remember I like to have really bright lighting! I also put in 4 zones—in other words I have 4 switches, each switch lighting up 7 of the lights. When all 28 lights are on, the arena has completely bright, even light, and is actually like riding in sunlight. It's really, really wonderful. There aren't any shadows or dark spots anywhere, and the horses LOVE it!

<u>**NOTE:**</u> You will not be able to have all the arena lighting on one circuit because it will overload the circuit. Plus you will want to be able to turn on different levels of lighting at different times for various activities in the arena and have the varied lighting levels yielding evenly spaced lighting. To be able to turn on different levels of lighting in the arena and have evenly spaced lighting means your various circuits should <u>**not** light just straight lines</u> of lighting fixtures. Each circuit should light the lighting fixtures in evenly spaced patterns of light across the entire arena. **(see Photo #67 and Diagram #4)**

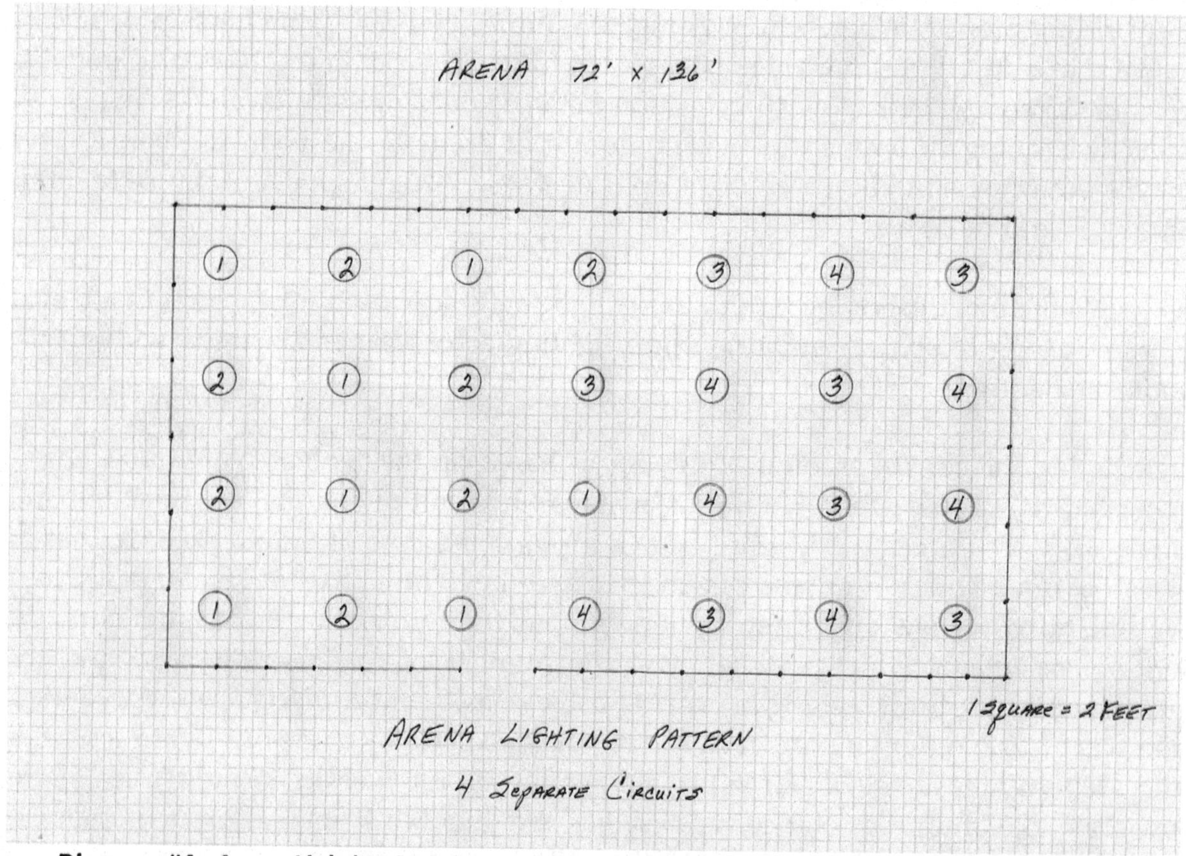

<u>***Diagram #4: Arena Lighting Pattern -- 4 Separate Circuits for 28 Lights – 7 Lights per Circuit***</u>

NOTE: There are a couple of differences between high bay and low bay lighting fixtures that should be mentioned. First of all, low bay lighting fixtures are used for ceiling heights (the height at which the light will be suspended) between approximately 12 and 15 feet in height. For ceiling heights greater than 15 feet, high bay lights are used. The advantages of low bay lights are: (a) they have lenses over the light bulbs which keep out the dust and therefore reduce the fire hazard; (b) the lenses evenly distribute the light making for a more evenly lit arena. My arena ceiling height (the height of the bottom of the trusses) is 16 feet, and the light fixtures are suspended down from the trusses so they are really at about 14 feet.

Photo #93: Exterior Light Fixtures Mounted on Wood Block

The downside of my having 76 foot candles of lighting is that it is very expensive to have all the lights on. So if you aren't going to be jumping, then you really don't need to turn on all the lights. What I do for turnouts and flatting the horses, is just turn on two sets of switches which means I have on 14 lights for a total of about 41 foot candles. The lighting is still even and without shadows or dark spots, it's just "less" bright lighting overall. If you are just turning out, you can get away with just turning on one set of switches, but there will definitely be dark areas in the arena. However, if your purpose is to get your horse to run and release his energy during the turnout, you might actually want to have some dark and scary areas in the arena to inspire him to spook a little and make him run!

Now let's get back to the barn itself and exterior lighting.

NOTE: All exterior light fixtures must be mounted on blocks of wood attached to the metal siding of the buildings. The size of the wood blocks will vary appropriately with the sizes of the various fixtures. The metal siding is not strong enough to support the weight of the lighting fixtures nor will the lighting fixtures hang correctly if mounted just to the metal siding.

Remember that it is dark in the country. There aren't any streetlights in the country and your neighbors are at too far of a distance from you for their porch or house lights to lend any significant light to your farm. So without lights—particularly on moonless or cloud covered nights--you are without any kind of nighttime visibility. I don't know how you feel about walking outside in the black of night in the middle of the country knowing full well that there's lots of wildlife—at a minimum—around you, but I know that it makes me feel very uneasy! As such, I would highly recommend that you consider lots of exterior lighting on both your house and your barn and arena when you live in the country!

That little old barn where my horse, Alex, stayed while he was recovering from his torn suspensory ligament had the greatest little **"pop on" light** (lights that turn on from a motion detection sensor integrated into the light) that would pop on just as I pulled up my car to park next to the barn. It fully lighted my car, the path to the barn, and the entry door into the barn. I found that light to be very reassuring. In fact, I loved that little barn's pop on light so much that having pop on lights was right up at the top of my dream farm list! And, I have to say, now that I have them on my farm, that they work out just great. They pop on "just in time" as I walk around the farm at night to make me feel safe, and then turn off after a little while so as not to waste any electricity.

It's a good idea to mount one double set of pop on lights (we used a pair of 150 watt each, halogen flood lights) at the top of every corner of your barn and arena, plus at the top of your barn end door eaves. You will want to have an equivalent set of lights on your hay barn and on each and every other structure you may choose to construct. In addition, you should light any nooks that might not be fully lighted by the lights you've installed up on the major corners of your barn and arena. You should also have pop on lights that light up your muck rack so that it's easy to see and access at night.

NOTE: Remember that pop on lights don't just pop on when you walk by. They also pop on when strangers walk by or when wildlife walks by. As a result, having lights that pop on strongly discourages wildlife (and strangers) from coming onto your farm.

Even with all this lighting, however, I still felt uneasy on the opposing side of my path around the barn from the pop on lights on the barn and arena structures. In other words, I couldn't see "out" across the pastures which left me feeling vulnerable.

Originally I had planned to have lighting run along the entire length of the driveway and to have "up lights" on each of the trees lining the driveway. But, once again, as you've heard too many times before, this didn't happen. The truth of the matter is that surprisingly, I'm glad now that I didn't line the whole driveway with "guiding" lights. That's because the flipside of living in the country, and being pretty much alone, is that you don't want to advertise exactly where you live or exactly how to approach your house and barn for John Q. Public. It's far better to have "just enough" lighting to feel safe and be safe, but not enough lighting to easily allow people to enter your property, especially at night.

So the way I solved the problem of not having—and eventually not wanting to have—full driveway lighting, was to select a couple of trees to light on the opposing side of the driveways from the barn and arena so that I have plenty of light on both sides of the driveway when I walk back and forth to the barns at night. On one tree—a pear tree—I keep white mini XMAS lights up year round, and on another tree—a large spruce tree—I have three up lights. I also have two up lights reflecting on the front door of the hay barn to light a sufficient part of the north end pastures to make me feel safe.

Speaking of the hay barn, let's not forget about the **hay barn lighting**. The hay barn needs an exterior light fixture by the door that I've already suggested should match the ones on the barn. The hay barn should also, as I've mentioned, have the same kind of pop on lights on all its exterior upper corners. There must be power outlets in the barn as well as at least one exterior outlet box. Inside the hay barn, you should have the same kind of enclosed, ceiling mounted fluorescent fixtures as in the barn aisle. Although it isn't necessary to have as much lighting in the hay barn as in the main barn, I do wish that I had put in more lighting than I did in the hay barn. In the 36 x 36 foot hay barn, I have three, enclosed, 4 lamp, fluorescent lights, but because they are mounted flush to the bottom of the trusses in the high ceiling of the hay barn, it reduces their effectiveness. I wish I had about 50% more lighting in the hay barn. Once again, I like things to be light and bright, especially when I don't know what critters are hiding in and under the hay bales. I also don't want to know. So, by having nice bright lights in the hay barn, the critters are generally discouraged from coming out of their hiding places while I'm in the hay barn with the lights on, and that works out well for me!

Photo #94: Barn Aisle Rope Lighting

NOTE: You don't want to suspend your lighting fixtures from the ceiling of the hay barn as you have in the barn aisle. The hay will be stacked up to the rafters (or trusses) of the hay barn and suspended lights would be in the way.

There is one other "fancy" thing you can consider doing inside, and even on the outside of your barn. You can line your barn aisles above the stalls with **rope lighting** which makes a great night light for your barn. Outside, you can also either line the porch roof with rope lighting or the peak of your barn or arena roof with rope lighting which would look very pretty. However, such exterior lighting would also draw attention to your facility which may not be a great idea as we discussed if you are a private facility.

NOTE: Don't forget to plan for lighting switches to turn the rope lighting on and off should you choose that lighting option.

Now how about a **video surveillance system**? This is something that is great to have. Remember way back when we were talking about having pastures on either side of the barn and how when you've finished looking at the activity in the pastures on the one end of the barn and then go to check on the activity in the pastures on the other end of the barn, that there will probably be something going wrong in the pastures you just checked? Well, just imagine if you had a surveillance display screen mounted in the aisle of your barn that was projecting images from surveillance cameras directed toward the pastures on both ends of your barn that you could just look at without having to walk up and down the aisle in either direction? Wouldn't that be great? How about when you were in your house or on vacation and you wanted to check on your horses to make sure the person you left in charge of the barn was doing her job and that everything was okay? How would you feel if all you had to do to check on everything was to pull out your lap top computer and hook into the surveillance cameras you have focused on your horses' stalls and the barn aisle and the pastures? Wouldn't that be great? Or how about if you saw truck headlights coming down your driveway late at night and all you had to do was look on your computer screen to see what was going on without having to leave your house to check on things? Wouldn't that be great?

These surveillance systems are affordable and easy to install. For example, a system with four cameras, a digital video recorder, and an internet connection for remote viewing can be purchased for about $1500 and easily installed. The only connections required are electrical power (plug-in outlets) and a CAT5 local area network cable run from the house to the barn that connects to the internet through a router. This would be one of the two telecommunications wires that you should run through your telecommunications pipe in a trench between your house and your barn (the other wire being your land line phone lines). Having this cable will also allow you to have internet connection for your computer in the barn.

NOTE: If you want telephone lines in your barn, remember that those lines must be run from your house to your barn through the telecommunications PVC pipe you installed early on in your construction.

So now we have discussed just about everything from a construction planning point of view. But there are a few more items we need to cover.

Chapter Nine

Farm Equipment and Final Landscaping

Let's return to the pastures and discuss a few more options to consider including in your pastures.

We had discussed how having treeless land is actually easier to deal with in the end, and decidedly cheaper than a heavily treed property because you won't have to pay to have any trees cleared. But now you're looking out at those treeless pastures and your farm in general and seeing the need for trees here and there for: (a) beauty; (b) sun protection for the horses to stand under in the summer; and (c) wind protection for the horses in the winter. This is where the "**big trees**" nurserymen and their "Big John" tree planting truck come in. But this is also where you will want to think about a couple of things regarding trees.

First of all, any tree that you put in a pasture **must** be surrounded by fencing to protect it from the horses eating the bark off it. Trees can't survive if too much of their bark is removed. Here it is best to use the PVC stiff plastic, four rail fencing. Each and every tree placed in the pasture should have an 8 foot square of PVC fencing placed around the tree which is centrally located within the 8 foot square.

Photo #95: Protective Fencing around Pasture Tree

Second, you should select deciduous trees (ones that lose their leaves in the winter) to plant as shade trees and trees that are not poisonous to horses. I planted Norway Maple trees in the pastures, and pear trees along the driveway.

NOTE: Don't forget to buy a book on plants, trees, and bushes that are poisonous to horses!

Third, to block winter winds, you should plant large evergreen trees such as blue spruce. You should strategically place the evergreen trees right outside the pasture fencing where horses can stand behind them (the horses would be inside the pasture, next to the fencing but behind the blocking pine trees) to block the cold, prevailing winds of winter.

If your horses will be turned out all day or all day and night, then you should build **run-in shelters** in your pastures. The run-in shelters will afford your horses a safe place from both weather and spooky things that make them want to temporarily hide. Remember, however, that run-in shelters are places that accumulate a LOT of horse waste. So if you aren't inclined to keep up after the cleaning out of your run-in shelters, and your horses aren't going to be turned out for any long periods of time, then you're probably better off just having trees for them.

A typical run-in shelter is 10 feet wide by 10 feet deep by 10 feet high, and three sided. However, this is just a minimum size. Your shelter should be big enough to accommodate all the horses in the pasture to be able to fit in at any one time, plus have extra room for a bossy horse. Also, the roof height should be high enough so that the horses won't hit their heads. The enclosed, back side of the shelter should back against the prevailing wind to take the chill out of the winter wind for your horses. Be sure that you don't put your shelter in a low spot where water accumulation will be a problem, and be sure that the shelter is both in a place, and of a size, where your tractor can easily access it and conveniently clean it out. The shelters would look best if they match your barn and arena with the same colors of metal siding and wainscoting on the exterior. The interior of the shelters should also be lined with #2 southern pine, tongue and groove lumber for strength, insulation, and durability.

So, now you have your whole farm pretty much envisioned. You can just see it all laid out so perfectly and beautifully in front of you. And it will be! But now you have to consider how you will keep your farm looking so perfect and beautiful!

In other words, you will need to consider what equipment you will need to operate and maintain your farm.

First of all, you should have a pick up truck and a horse trailer. Even if you were planning to use someone else's truck and trailer to transport your horses on that first trip to your brand new farm, you really need to have your own truck and trailer. You just can't predict when one of your horses might get sick or injured and need to be trailered either to a central veterinary facility or to a teaching university veterinary clinic. Remember that there's a 50/50 chance that the horse will get sick or injured at night and need to be immediately transported. Trying to rely on someone else to transport your horses for you does not generally work out very well. So, it's really best to have your own equipment.

If you are buying a new pick up truck and horse trailer, I would like to point out a couple of things for you to keep in mind.

First of all, let's discuss a **horse trailer**. I would like to recommend that you get an all-aluminum horse trailer in that it is much lighter to pull and less inclined to rust. You should buy a trailer that has nice big, opening windows and window grills, plus an openable, roof top air vent in order for the horses to have plenty of air flow. You should line the base of your trailer with thick, rubber mats if they are not already included with your trailer purchase. Larger wheels on trailers (16" wheels) make for a smoother ride, and radial tires on your trailer will give you better traction.

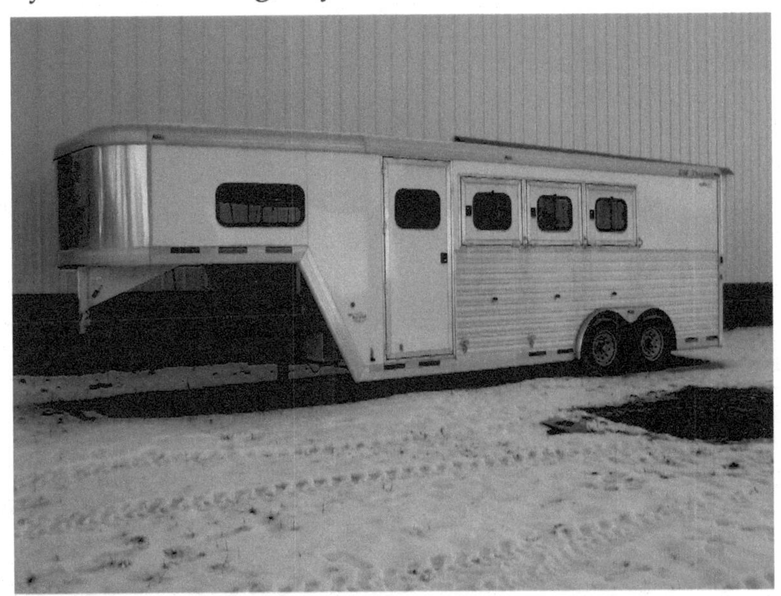

Photo #96: Goose Neck, Slant Load, Three Horse, Aluminum Trailer and Tack Room

Beyond that basic information, you should know that there are bumper pull trailers, and there are goose neck trailers. Although I don't have a bumper pull trailer, I have had some limited experience with them that made me not choose for them. Bumper pull trailers are only good for one or two horses, and they are straight loading which is never happy for horses—even worse for exiting out backwards than for loading in frontwards. Also, the tongue load on the bumper makes the weight distribution unlevel in the horse trailer which means the horses are standing on an unlevel platform while being transported which is not good for their legs. In addition, the bumper hitch connection is not a very strong one, and has been known to disconnect during travel. While traveling, any big accelerations or quick stops that occur will cause the horses to shift front to back or back to front, and because they don't have any support wall to brace a major portion of their bodies against, their legs will have to take the full burden of balancing their weight which is very hard on their legs.

Photo #97: Goose Neck Connector

Having a bumper pull trailer disconnect from the truck actually happened to a friend of mine while transporting his horses to a horse show. I found out about the situation when our mutual trainer suddenly announced at a horse show that she had to immediately leave the show to help this man collect his horses and transport them to the show. She said that his bumper hitch had detached while he was driving to the show down the interstate, the horse trailer had rolled backwards down a hill into the median, and the back door of the trailer had been damaged when the trailer came to a stop into a road sign, such that the horses had gotten loose and were unrestrained on the interstate.

Hearing that situation was my final determiner to never have a bumper pull trailer.

I chose for a goose neck, slant load, three horse, aluminum trailer with a tack room. Loaded, this trailer has a gross vehicle weight of about 10,000 pounds. It has a double axle with electric brakes.

A goose neck trailer has a much stronger hitch connection than a bumper pull, and is more easily maneuverable to drive and park than a bumper pull, especially when backing up. The goose neck hitch hookup hardware is located in the bed of the pick up truck. Although it's a little more difficult to align the truck bed with the goose neck of the trailer to hook it up, it's worth it for the maneuverability and the strength of the connection. Also, it is worth it to be able to have a slant load trailer for the horses. Slant load allows the horse to have the full width of the trailer in which to load as well as enabling the horse to turn around and face the back of the trailer when exiting the trailer--all of which causes

Photo #98: Full Rear View of Open Horse Trailer showing Slant Load with Dividers and Separate Loading Ramp

Photo #99: Separate Loading Ramp in "Up" Position

much less tension in the horse regarding loading and unloading. While loaded, the horse is standing at approximately a 33 degree angle to the front of the trailer with a full length support wall for him to lean against so that his balance and legs are much more able to handle any quick accelerations or stops.

Photo #100: Separate Loading Ramp in "Down" Position

NOTE: Please be sure to have a separate loading ramp (**not** a ramp that forms part of your actual rear trailer door system) installed on the back of your trailer. You will greatly increase your loading and unloading safety and success with a separate loading ramp. After closing the rear trailer doors, this ramp lifts up and locks into position.

Photo #101: Separate Loading Ramp in "Down" Position with Opened Rear Doors

For some reason, most slant load horse trailer manufacturers these days make a trailer that is totally dysfunctional in my opinion. As I just mentioned, the reason I like a slant load trailer so much (with a loading ramp) is that the horses have the full open width of the trailer to use to load and unload, plus a full, unrestricted view of the whole interior of the trailer when they are loading—all of which makes for a calm horse. So to my mind, seeing trailer manufacturers place a wall across half of the back of the trailer, and then a floor to ceiling saddle rack in front of the wall, seems totally counterproductive to one of the major reasons for having a slant load trailer.

Another amazing thing the slant load trailer manufacturers don't do is they don't put a locking, restraining wall for the last horse in the trailer to brace against while traveling, but just leave an open, pie shaped area for the last horse in the trailer to stand in and be sloshed around in as the trailer moves forward, stops or turns. In other words, a three horse trailer only has two restraining wall slots for two of the horses with the "third" slot only being that open, unrestrained, pie shaped area. So in my opinion, what they are calling a 3 horse trailer, is actually a 2 horse trailer. What are they thinking here?

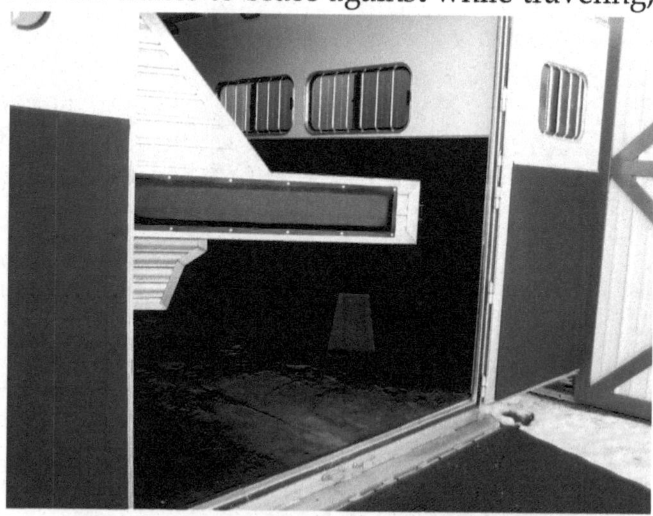

Photo #102: Custom Third Restraining Wall with Locking System

So what did I do to solve what I saw to be counter productive problems with standard slant load horse trailers so that I would have one that I felt was correct? I found a horse trailer manufacturer who was willing to work with me to custom design the modifications I wanted, and at a reasonable cost.

Photo #103: Saddle Rack Installed in Trailer Tack Room versus Rear of Horse Trailer

First, I moved the saddle rack from the back of the trailer into the trailer's tack room which is a separate room located in the front of the trailer that has plenty of room for a saddle rack. Next I had the manufacturer completely remove the swinging wall from across the back of the trailer so that the trailer once again was opened up with an unrestricted view. Then, I lengthened the trailer by about 1 and ½ feet, just long enough to accommodate a last horse support wall and allow the wall to snap into a secure fastener like the other separator walls between the other horses. So now I had a real, three horse trailer with three, securely locking separator/stabilizer walls for all three horses.

The other important thing to know about horse trailers is that one size does NOT fit all. When manufacturers talk about their standard trailers, they are talking about horse trailers that are made for the average sized horse—more like a 15 hand, quarter horse. If, however, you have large, warm blood horses that stand 16 to 18 hands and have wide girths, having a standard sized trailer just isn't going to work. You may be able to get the big horses into a standard sized trailer for one trip, but good luck ever getting them to scrunch back into an inappropriately small trailer the next time you need to ship them!

A standard sized trailer is 6 to 7 feet wide and 6½ to 7 feet high. What is available to you is having an 8′ wide by 7′ 7″ high trailer. If you have big horses, then you will need to purchase a correspondingly big sized trailer.

NOTE: Really big horses, like over sized warmbloods and draught horses, need an entirely different kind of horse trailer to accommodate their size. There are center ramp loading, tractor trailer type, horse trailers for these types of oversized horses.

And, finally, the big danger for us humans with horse trailers is the manual lift crank that has to be turned by hand in order to "lift" the front end of the trailer high enough to connect the trailer hitch to the hitch hardware in the bed of the truck. Everything goes just fine with such a manual crank unless or until things don't go just fine due to such things as your strength running out, your concentration being distracted, etc., such that you let go of the handle. Then, suddenly, the crank turns into a very efficient, finger removal system as it spins rapidly out of control. The crank will also be happy to remove or injure your hands or any other body parts, such as your head, that get in the way of its wild spinning.

Photo #104: Battery Operated, Electro-Hydraulic Trailer Lift

But there is a safe way around such a potentially atrocious situation, and that is to spend the extra money to purchase a battery operated, electro-hydraulic lift for your horse trailer that lifts your trailer up to your hitch with just the touch of a button and without the worry of any wildly circling, heavy metal parts. Needless to say, I am once again strongly urging you to spend the extra money to get an electro-hydraulic lift and reduce as much risk of injury as possible during the trailer hitching process.

After researching trailers, and coming to realize that horse trailers actually come in sizes, and that there were trailers made specifically for horses, I became really upset about the trailer our boarding barn had used to haul all the farm's show horses back and forth to horse shows. Their trailer never looked quite right to me. What was especially of concern to me was the lack of air I felt had to exist in the trailer in that it had NO windows, but only a couple of thin rows of air slats that ran along the top of the long sides of the trailer. I would just stare in horror at our horses jammed into the windowless trailer on scorching hot days and wonder at how they were going to survive the trip to the show or back home.

Photo #105: Horse Trailer showing One Opened Window with Grill in Place

Well, I was right about it. Horses DO need to have the big windows and a LOT of air flow or their health and survival are at risk while being transported. Also, what I came to find out was that our show horses were being hauled to horse shows by that boarding barn in what is a "stock" trailer meant to transport cattle—therefore the air slats and no windows. In addition, for my own sanity, I walked inside the boarding farm's trailer to see if I could find any telltale signs that the trailer was way too low for the farm's many 16.2 hand and larger, warmblood show horses. Sure enough, I found multiple dents in the ceiling of the trailer where the horses had smashed their heads into the too-low ceilings. There were also plenty of hoof dents along the sides of the trailer where you could see the horses had been kicking. What I now realized was that the reason I'd heard the loaded horses kicking was from frustration over being squashed into a trailer that was way too narrow and low for their huge bodies.

Photo #106: Horse Trailer showing One Opened Window plus Lowered Window Grill

NOTE: Horses travelling great distances to shows which require the horses being in the trailer for more than a few hours will also have feed and water requirements that need to be addressed for the horses' health and well being.

Now let's talk about pick up trucks. We've all passed pick up trucks and have seen the big, chrome, numbers shining brightly into our faces off the front end of the truck. Talk to any trucker and he'll be proud to tell you he's got a F350, GMC 3500 HD or a RAM 3500 one-ton truck as compared to your little F150, GMC 1500 or half-ton truck. Yet, ask anyone—be it a truck owner or a truck dealer—what those numbers actually mean that are bolted onto their truck, and what you're going to get back is either a statement that it means the engine is bigger in their truck because the number is bigger on the side of their truck or they will tell you that it has something to do with the load carrying capacity.

But, why, you ask, do you even need to be concerned with those truck size numbers? Because there are several factors that need to be taken into consideration in order to have the correct truck to pull your horse trailer—load capacity, pulling or towing capacity, engine size, and engine torque.

First you need to understand that the number on the front of the truck has to do with the amount of weight that the truck can hold (carry) in the truck bed. The bigger the number, the more weight the truck bed can hold. For example, a GMC Sierra 1500 can carry about 1800 pounds of rocks in its bed, whereas a GMC Sierra 2500 HD can carry about 3500 pounds of rocks in its bed.

In regard to towing a horse trailer, the significance of that number is to have enough weight load capacity in the truck bed to take the weight (tongue load) of the goose neck trailer connection into the bed of your truck—which can vary by significantly large numbers. The weight of your goose neck connection in your truck bed is calculated by using a percentage of the weight of your trailer added to the weight of all your horses loaded into the trailer, plus the weight of any additional equipment and supplies that you intend to load into your trailer. As an example, if your trailer has a loaded, gross vehicle weight of 10,000 pounds, you can figure about 15% of the gross vehicle weight will be distributed to the goose neck (tongue load) in the truck bed. That means to handle that trailer's tongue load weight, your truck must be able to carry 1500 pounds of weight in its bed.

The second issue with a pickup truck is how much weight the truck can pull. That is a different calculation than how much weight the truck can carry in its bed. How much weight a truck can pull or tow has to do with the engine size of the truck plus its power and torque.

In our case, we have an aluminum, three horse trailer with tack room, which when fully loaded weighs about 10,000 pounds. With a 6 liter V8 gas engine that gave the truck a trailer towing capacity of 10,000 pounds, we were able to pull our horse trailer with a 1500 truck, but our truck was pretty much maxed out.

So if you are planning to pull a 6 horse trailer with 16+ hand warmbloods, then you would need to get a truck with a more powerful diesel engine, and one that is designated as heavy duty.

By the way, the number on the truck has NOTHING to do with the engine size of the truck—just the load capacity of the bed! But as I mentioned, you may indeed need a truck with a bigger engine size as well as a bigger bed load capacity if you are pulling a really big horse trailer. Just be aware that there are many factors that go into determining what truck you should buy—basically pulling power plus bed load capacity.

After a lot of phone calls, and research into truck manufacturer's websites that resulted in a total lack of information, my husband finally decoded the mystery of the meaning of those chrome "engine" numbers bolted (or glued) on the front of all those pick up trucks. In summary, here's what it means.

Light duty pickup trucks generally are offered in three load ratings which are labeled as 150, 250, and 350 or as 1500, 2500, and 3500. These load ratings are sometimes called ½ ton, ¾ ton, and 1 ton ratings. It is actually better to think of these numbers as the approximate load capacity (15% of your trailer's gross vehicle weight) in the bed of the truck which would be for a 1500 pound, 2500 pound, and 3500 pound bed load weight carrying capability, respectively. However, you should determine what the actual load ratings are for the truck you want to use thru the dealer/manufacturer because some trucks, though labeled as 150's for example, may have a higher bed load capacity than the 150 would indicate.

<u>**NOTE:**</u> You need to be sure and double check these basic ratings remain valid for a specific truck in that other factors such as options loaded onto a basic truck can reduce the load capacity of the standard ratings. For example, some "designer" trucks have a lot of heavy chrome detailing, accessory equipment, interior trim, truck bed covers, and the like, added to them which adds a lot of weight to the basic truck itself and thereby greatly reduces its bed load capacity.

<u>**NOTE:**</u> The limiting factor in non-commercial trailer towing (about a six horse trailer or smaller) is a combination of engine size (power and torque), and the chassis load rating.

The next issue regarding pick up trucks has to do with the bed length. During one of the conversations we had with our horse trailer supplier, she happened to warn us that if we had a short bed truck that we must be careful not to turn too sharply or the top of the front end of the horse trailer would crash into the back window of the pick up truck.

What????? We panicked. Why did we panic? Because we knew that we hadn't bought a long bed truck and therefore we assumed we had bought a short bed one. But, to our relief, upon investigation, we were relieved to find out that there is such a thing as a "regular" or "standard" sized truck bed which is what we had.

<u>LESSON #54 :</u> Be careful about the bed length in your truck if you intend to pull a goose neck horse trailer.

So after we had determined that we were going to buy a goose neck horse trailer, we had to have the goose neck trailer hitch installed in the bed of our pick up truck.

LESSON #55: You can't install a goose neck trailer hitch in the bed of all pick up trucks. Some pick up trucks have "light duty" beds that don't have enough strength to attach and operate a goose neck in them.

Dumb luck once again, we had purchased a pick up truck with a strong bed and could install a goose neck trailer hitch. Now the next "problem" was preventing the metal bed in the pickup truck from rusting out.

Most people lay a plastic bed liner in their trucks as a rust preventative for their truck beds. However, the problem with removable liners is that they don't fit flush and therefore water gets under the liner and eventually rusts out the bed of the truck anyway. The solution to this problem is to plan ahead, **before you have the trailer hitch installed**, and take your truck to a shop where they do "spray on" rubberized composite, truck bed "liners." Spray on liners totally seal the metal of your bed and keep it from rusting. Of course if you manage to scrape off the liner in places, you will still get rust in those areas.

So what other equipment do you need to maintain your farm? Below I will try and give you a feel for the minimal amount of equipment—both light weight and motorized—that you will need.

For mucking out your stalls and keeping your aisles clean, you will need a wheelbarrow, a muck rake, a broom, vacuum cleaners, and a dust pan.

- You should buy the biggest capacity, two wheeled, easy dump wheelbarrow you can find. There are horse barn specific wheelbarrows for sale in many horse supply catalogs and online.

- For a muck rake, you should get the lightest weight aluminum muck rake possible as well as one that is ergonomically designed. Having the proper type of muck rake will save your hands and back a lot of wear and tear.

- The best broom to have is a stiff, corn broom which you can buy at just about any store. Don't buy short bristled or soft bristled nylon brooms. I also personally don't care much for push brooms. Push brooms are "okay" for really long aisled barns for quickly sweeping the middle of the aisle, but they are clumsy and make a LOT of dust. Plus you will need to also have a regular corn broom anyway to get into the nooks and crannies along the stall fronts of the aisles.

- Your dust pan should have the largest possible capacity and have an upright handle of at least 3 feet so you don't have to bend over to use it.

- Having vacuum cleaners is also handy. First I would recommend having a regular "house type" canister vacuum cleaner with disposable bags for removing the general dust in the barn because a shop vacuum is clumsy to use, doesn't have enough suction to pick up fine dust, and is messy to empty out. But, you also need to have a wet/dry shop vacuum to pick up big chunks of debris in the barn, and especially for those times when you need to suck up water as we've had to do when the stalls flooded.

- You should also have a good supply of dust rags, buckets and sponges, a wet mop, and a long poled cob web catcher, plus the regular array of cleaning solvents.

Next there is the motorized equipment you will need.

First and foremost, you will need a tractor. The smallest tractor size I would suggest is a **compact utility tractor** with a diesel engine rated at about 35 horse power plus a **rear PTO** (power take off) which is used to power such things as your cutter. The tractor should have at a minimum as front end attachments, a **front bucket loader** about 5 feet in width, and a fork lift attachment.

LESSON #56: If your tractor can't lift enough weight or lift the weight high enough, you may find there are certain chores on your farm that you will be unable to do with your tractor. Your attachments must be appropriately sized for your chores.

There are some issues that I have encountered regarding the bucket loader, and the fork lift attachments about which I will give you a heads up. First I should note that you will need to think beyond what I present here as you consider your specific needs for your farm and customize your tractor for those needs. Also, I should note that there is little information online that allows you to collect the necessary information to make such decisions. You will need to go to your tractor dealer and discuss with him the specifications of his equipment and its attachments, including all weight, height, and loading limitations, and any other considerations you will need in order to accomplish your desired chores on your farm.

For example:

(1) If you are going to have commercial truck removal of your muck rack waste, that will most likely necessitate your tractor transferring the muck rack waste into a dumpster. In order for you to be able to do such a transfer, you must be sure that your bucket loader is smaller than the opening of the top of the dumpster container so that

no muck will fall out of the sides of the bucket loader while being transferred into the dumpster. In our case, we have a 6 yard waste container that is 6 and ½ feet tall and 6 feet wide. However, the dimension of the actual hole in the top opening of the container is 5 and ½ feet (66 inches) wide, and therefore the bucket loader cannot exceed that 66 inch wide opening, and realistically must be somewhat smaller. We have what is called a 5 foot bucket loader. The actual width of the outside of our 5 foot bucket loader is 63.5 inches or 5 feet plus 3 and ½ inches. The interior of the 5 foot bucket measures 5 feet (60 inches). As you can see, our bucket loader fits inside the top opening of our dumpster to allow for aligned dumping of the muck.

The other factor that must be taken into consideration is how high the tractor can lift the bucket. Remember that the dumpster container is 6 and ½ feet high. So your tractor has to be able to lift the bucket loader that high in order for it to be able to dump into the container. Our tractor is just able to lift high enough to dump. When we purchased our tractor we had two lifting height options and we selected the "higher" lifting option. In summary, you must make sure that your bucket loader is not too wide for your dumpster container, and that your bucket loader can be lifted high enough to dump into the dumpster container.

(2) The fork lift attachment is used for several chores around the farm. However, here too, as with the bucket loader, you must make sure that certain weight and height specifications are matched to your chore needs. For example, if you are going to use your fork lift to stack or remove hay bales from tall stacks of hay, then you have to make sure that your tractor can lift up as high as you need, and that the tractor can support the weight that it is lifting up to, or down from, your specified height. In other words, just because a tractor can lift high, it does not mean that the tractor is heavy enough to lift any significant amount of weight without flipping over. Also, as previously discussed, if you wanted to buy bagged shavings in bulk from a tractor trailer load, then you would need to offload the bagged shavings from a tractor trailer with your fork lift. In order to offload the shavings, you would not only need specific lifting height ability, but you would also need a large tractor weight lifting capacity in order to be able to remove such heavy pallets from the tractor trailer. A wood pallet, plus 45 bags of shavings will weigh approximately one thousand pounds per pallet. So the ability to use a fork lift attachment on your tractor is limited by the weight your tractor can lift, and how high your tractor can lift that weight.

Maintenance attachments for the tractor to pull would include, at a minimum, a **rotary cutter** to rough-cut mow your pastures, a **landscape rake** to top dress your soil and your driveways, a **box blade** to scrape, redistribute, and level your driveway stones and stone dust, and better, a one ton roller attachment to level out your pastures and fields in the spring when they have become uneven from the winter freezes, thaws, and snows. The tractor we purchased and use has a **quick attach/disconnect adaptor** that makes it easy to connect and disconnect attachments to the three point hitch. Having the quick attach/disconnect adaptor is enormously convenient because, during a typical day, you frequently have to use multiple attachments.

You should also purchase a **four wheel drive, Gator type vehicle** with a dumping back end (automatic lift/dump bed), a heated cab with a windshield wiper, a trailer hitch on the back end, and a snow plow blade attached to the front end (if you live where it snows!). This vehicle will be used to do all kinds of tasks around the farm from delivering the hay from the hay barn to the horse barn, to dumping uneaten stall hay out in a dedicated corner of your property (where the deer will be happy to consume it for you!), to carrying your landscape supplies around the farm, to pulling the arena drag or rake behind it, and then also to plowing the snow to clear the driveways and pastures in the winter. You basically cannot run a farm without a Gator!

Next you will have to have a **zero turn lawn mower** with at least a 54" cutting blade ability. Not only will you use your mower to mow the parts of your farm that are considered lawn areas, but it will be used to "rescue" pasture grasses that have grown too long as we discussed earlier. You will also need a small "push mower" to mow the small areas that the zero turn lawn mower cannot access.

You will also need either an **arena drag**, at a minimum, or an arena rake to drag your arena. The drag or rake can be pulled behind the Gator. (Remember that the arena rake should be wider than the Gator!) If you can afford to also have an ATV, it would be easier for you to regularly drag your arena using the drag or rake pulled behind an ATV.

Another great thing to buy, but not absolutely necessary is a pull behind, **fence post edge trimmer**. When you have big pastures with a lot of fencing and fence posts, there's a lot of edge trimming to have to do with a hand held grass trimmer. In fact, your back will most likely give out long before you've finished edging just one pasture! As such, buying this gas powered, pull behind fence post edge trimmer (which you would pull behind your Gator) is a back saving smart purchase.

In addition to having a **hand held fence trimmer and a leaf blower**, you might also want to have a **power washer and a gas powered cultivator.** I would also highly recommend having a mobile, free standing, gasoline powered **generator** if you haven't installed an automatic, gas powered, switch over generator in your basic farm plan in case you have a total power failure that lasts for any significant period of time.

NOTE: The mobile, free standing, gasoline powered generator will need an electrician installed, transfer switch or a "reverse feed" connection in order for it to be functional.

Lastly, it is nice to have an **enclosed utility trailer**. These pull-behind trailers will serve you in myriad ways including everything from the transport of shrubbery and farm supplies to the removal and disposal of moldy hay bales from your hay barn!

Well, there you pretty much have it.

As you can see, building a horse farm is quite a journey. Once I had completed my journey, I have to admit to you that I wasn't quite sure if I was ready for the funny farm or if I was living on it! I wish I could be there with you when you finish your journey so that we could compare notes. But, alas, I can't. All I can do is hope that you found a wealth of helpful information in this book to help you and your horses build the best horse farm ever! Best wishes to you for unending happiness.

Chapter Ten

Epilogue

So what's it like once you finish your dream farm and move in your horses to their new, perfect barn?

Well, there you are, finally driving down that nice, long, solid driveway in your properly rated pick up truck, past all those beautiful rows of white horse fencing, and fields of rich, green pasture grass, pulling your horses behind you in their perfect horse trailer. As you pull to a stop in front of your beautiful brand new barn, and prepare to unload your horses, you can hardly contain your excitement. Just think of how much the horses are going to love this place! After all, haven't you planned absolutely everything from their point of view and for their health, happiness, and comfort? You just can't stop thinking about how ecstatic your horses are going to be when you open that trailer door and they see all this wonderfulness spread out before them, and for them! You are sure that they will be the happiest horses in the whole wide world! How couldn't they be? Why wouldn't they be?

So you contain all your excitement so as not to overly excite your horses, and force yourself to be calm as you unlatch the loading ramp, and then open the back of the trailer. You proudly unload the horses, one at a time, and lead them into their new home. The horses are clearly excited. After all, they most likely think they are at a horse show and will soon have demands placed upon them to perform. But how happy they will be when they find that, no, there is no demand here on them—this is their new home—their perfect home—the place where they can totally relax and find perfect peace. You now feel confident in your heart that all the time, worry, planning, money, and stress that you have been through on the project has been well worth it to live this moment when your dream farm reality has finally come true!

But now that you have all the horses in their stalls, and have given them a significant amount of time to settle down, you find that they aren't settling down. In fact, what you find is that your horses seem to have become even more unsettled than they were when you first unloaded them.

What's going on? What's wrong? What in the world could possibly be wrong? Everything is perfect, isn't it? Haven't you made sure it is? Can't the horses see how wonderful it all is? Aren't they just thrilled to be in such big, spacious, airy stalls with views to look out of in all directions?

Well, they should be, but somehow they clearly aren't!

And then worse, you watch as your horses go over to the water bowls on their automatic waterers to get a drink and BWAM! When the waterer starts to refill, the horses panic and jump back flat against the opposite stall wall from the waterer and shake in fear! They look at you as if to ask what kind of cruel joke you are pulling on them not allowing them to even get a drink of water without getting the daylights scared out of them in this spooky place you've brought them to?

And now you feel absolutely HORRIBLE! You feel like everything you've been through, everything you have done, has been a total waste of time and money and straight out WRONG! You instantly conclude that after all this, your horses are clearly totally MISERABLE and that your dream farm is just a nightmare!

This is most likely what will happen when your horses move into your new farm as it did for me. I was completely devastated. However, I didn't have this book to read, so I felt truly AWFUL, and had no where to turn but ON myself. I decided that I had somehow totally and miserably failed.

Yet, logically, I thought to myself, how in the world could my horses have been happier at that dark, dank, airless, hole of a barn covered in filth where they had lived for so long? That just didn't make any sense.

As I repeated that logic flow to myself, like a mantra, over and over again, my mind finally latched onto the "filth" word. That's when part of the problem that was happening occurred to me.

The new barn was just TOO new, and TOO clean for the horses.

Why do I say that?

Well, first of all there were no "horse" smells in the barn. So to a horse's mind, that meant that this was NOT a place for horses. Also in their horse brains that meant they were not in a "safe" place for horses.

Potentially worse for the horses' minds was that there was just one place they had ever been where there weren't any readily detectable horse smells, even though there may have been other horses present.

And where was that?

The veterinary clinic. You know, that place where owners take their horses to get prodded and poked, injected and medicated, cut and sutured or wrapped and returned? That is not a happy place for horses to remember being.

So based on the lack of horse smells, and their personal experiences in such places as veterinary clinics, my horses felt pretty sure that I had taken them to a place where they should be on guard.

But rest assured that this panic my horses had, and your horses will most probably have, does go away in time when their own horse smells start to take over every inch of the barn. But it takes a few days.

In addition, veterinary clinics have brightly lit, spacious stalls. So the new bigger sized, brightly lit stalls and barn just served to further confirm to my horses that they were in the wrong place. Plus they had been used to living in stuffy small stalls where they didn't have to make any decisions as to where to stand, where to urinate, where to defecate and where to eat. There was no choice because there was no space.

In time, you will find that your horses, like mine, will come to LOVE and value their spacious stalls, the choices, the views, and the bright lighting—both natural and artificial.

Now, remember, you can bring in some buckets of their soiled stall shavings from their current facility and spread the shavings with all their smells on the new shavings you have put in their new stalls in your new barn before they arrive. However, you'll probably find that hard to do considering how pretty you'll want to keep your dream farm looking. But the truth is that your dream of keeping the farm clean will be short lived anyway. So, if you can bring yourself to bring in some soiled shavings, it should make your horses' transition into their new stalls go a lot smoother!

NOTE: If you are not going to be using EXACTLY the same grain/feed at your barn as your horses were being fed at their old barn, then you must be sure to bring along enough grain from the old barn to gradually transition your horses (over about a week of feedings) to eating their new grain. You need to gradually add more and more of the new grain to the daily feedings as you gradually reduce the old grain from the old barn, starting with as little as $1/5^{th}$ new grain, to $4/5^{th}$ old grain, for the first feedings. This is also true for hay.

For horses like mine who were used to living in a barn with a LOT of other horses, suddenly living in a small barn with just a couple of friends seemed lonely and spooky and made them feel insecure. There is a definite big barn versus small barn mentality in horses and they actually do feel more secure with a great number of horses around them. That's just natural in that herds in the wild are generally larger than just 3 or 4 horses.

But in time, your horses will get used to being in a small herd.

NOTE: Buy a radio for the barn and centrally locate it by the stalls. Turn the radio on as the first thing you do in the morning, and then turn the radio off as the last thing you do at night. Your radio gives your barn all the "noise" your horses will need. Horses enjoy quiet, calm music, like soft jazz or classical music, which keeps them mellow. No kidding! There have actually been studies done on this!

Then there's the automatic waterer issue. At first, to the horses, the filling cycle is loud and scary.

But in time, they learn to use the waterers and truly LOVE the automatic waterers.

I have two geldings and one mare. It was the mare who was daring enough to tough it out through the noise and action of the refilling bowl and figure out that she was actually in continual water source heaven. Through embarrassment of having the mare clearly able to stick it out and drink through the refilling noise, the geldings finally toughed it out as well. I did periodically supplement the automatic waterers the first night and day with a water bucket on the floor. But I was careful not to overdo it with the supplementation in order to encourage the horses to learn to use their new, wonderful automatic waterers.

NOTE: A good way to introduce the horses to the filling noise of the automatic waterers is for you to enter the stall and clean out the water bowl in front of the horse. Take off the rim of the waterer, remove the water bowl, dump out the water in the wash stall drain, put the bowl back in the stand, let it fill, put on the rim. When the horses see you unafraid of the bowl and its accompanying noise, and understand that the noise is just part of the watering system, they will come to understand that it is okay to drink from the bowls. Just a couple of dumps and refills with an hour or so in between the "dumps," should do the trick!

So, horses being horses, which means they are prey animals, it is only natural for them to at first be afraid of everything and anything that is new. But also, with horses being horses, they come to adjust to patterns in time and eventually find comfort in the repetitive pattern of any safe and healthy new pattern offered to them.

So fear not, your horses will come to LOVE EVERYTHING about their new farm—your dream farm—that they will come to know you built for them out of care and love and also come to thank you for providing for them every minute of every day that they are blessed to live with you. And, they will let you know their gratitude in myriad ways!

In the end, I think you will come to find in yourself a new level of pride and separation between you and those who "just" own, ride, and show a horse out of a boarding facility. Once you have 24/7 care of your horse/s you will become in time what is called a horseman. Once a horseman, you will find a new level of deeper relationship with, and understanding of, your horses. For those of you who have been raised on a horse farm and have known horses before you rode, you already know what I mean. For those of you who build your horse farm after years of ownership, riding, and showing, you will come to understand how much better a rider you would have been had you had your horses on your own farm first, and then rode and showed. You will also come to realize how much your former trainer and/or boarding facility owner was trying to keep you ignorant about horses and horsemanship in order to keep you dependent upon her/them for her/their personal profit. Remember that there are many people who can call themselves equestrians, but few who can call themselves true horseman!

Horses truly are the most wonderful creatures in the world for humans. There are endless lessons to be learned from the quiet dignity of these ever loving and forgiving, noble beasts that God has given to those of us who have chosen to be with horses. Whether or not you ever actually choose to build a farm, just the fact of your having chosen to be with horses, is one of God's greatest blessings to you. Best Wishes!

THE END

POEM
On Building Cloverleaf Farm

When I lived by the ocean,
As I stood on the shore,
I needed to sail to the place
Where the sky touched the sea;

When I lived in a valley,
Where the long road began,
I needed to enter its path and travel to the point
Where the two edges narrowed into one;

When I lived in a darkened woodland,
Next to a flowing stream,
I needed to follow the relentless flow,
To the place where gravity no longer pulled;

Yet, as I live in the openness of green fields,
Surrounded by unlimited, unobstructed views,
I feel no need but to embrace all that is above, around, and under me,
For here I have found the peace of all I love being one with me.

Southwestern United States

The southwestern desert region of the United States is one region where there is a significant difference in the approach to horse barns and horse welfare that merits discussion. I would like to share with you some observations from my personal experience boarding my horses in the southwest. Once again, these observations are presented as a starting point for your thinking and research if you are planning to build a farm—which is called a ranch in the southwest—for your horses in the desert region. The observations are not intended to be a complete list of issues that should be considered and addressed, but merely ones that I have noticed while boarding, and ones that I felt were important to share.

Basically, the horse's needs for seasonal adjustment in the southwest are a complete flip of those in northern, four season climates. The survival months for horses in cold weather climes are essentially the winter months of December, January and February when unprotected horses can succumb to subzero Fahrenheit temperatures. In the desert, it is the exact opposite with the summer months of June, July and August being when unprotected horses struggle for survival with daytime Fahrenheit temperatures that are regularly in the triple digits, and frequently exceed the 110 degree mark. So while the horse owners in the north start planning out and worrying in the summer months about how many blankets they'll need to keep their horses warm in the winter, the south westerners start worrying in the winter months about how they're going to keep their horses cool and shaded in the summer!

Photo #107: 12' x 12' Open Stall

Although there are enclosed barns in the southwest, the predominant style of barn is to have open stalls, and, in my opinion, that is the best overall housing for horses in the desert. I do want to clarify here that I do not personally feel that completely open stalls are a good idea although they are common and popular with backyard horse owners. When I talk about open stalls, I mean stalls that have half-kick-wall/half-tubing-wall stall fronts, and just open tubing "walls" for the rear side of the stalls, plus solid, kick-proof, separator walls between the stalls. Although the need for free flowing air into and out of the stalls is vitally important in the southwest, there are other weather conditions such as dust storms, wind storms, and torrential rains when a horse needs to have a solid wall of protection that he can stand behind. There is an additional need for protection which I will discuss below.

During the desert winter, the lowest nighttime temperatures are generally in the mid 30's. Rarely does a nighttime temperature dip into the 20's and even more rarely does the temperature dip that low for more than two nights in a row. Blanketing a horse for warmth at those temperatures for their open stalls is easily achieved even with a single blanket. The horse's natural winter coat growth easily protects any exposed body parts of the horse such as face, neck, stomach, and legs without the need for any extra protection.

During those same winter months in the desert, the daytime temperatures average in the high 40's/low 50's and even into the 60's. Such moderate daytime temperatures require that the horses' blankets be removed in the winter during the daytime hours. So, as you can see, there is very little need to protect the horses from "cold" weather, and, therefore, an open stall is easily tolerated by the horses in the cold desert months.

The rest of the year in the desert, the need for heat relief and fresh air is the more critical issue to be addressed for horse housing, and the basis that justifies the preference for open stalls.

Photo #108: 12' x 24' Open Stall

The rules for actual size of stalls remains the same in the southwest with a 12' x 12' stall being the preferred minimal standard for a full sized horse. The major expense for open stalls is in the roofing and the solid, separator walls of the barns. The tubing that provides containment stall "walls" is relatively inexpensive. As such, stalls in the desert are often extended twelve feet out beyond the barn roof making 12' x 24' stalls common. This allows the horse more space within his stall and also allows him the choice of standing under the protection of a roof or out under open sky.

Although these oversized stalls seem "perfect" for a horse, there are some drawbacks that should be pointed out.

First, there are only tubing "walls' separating your horse from the horse stalled next to him (or on both sides of him) when both horses (or all three horses) have chosen to be out under the open sky. Whereas your horse might be a perfect gentleman (or lady), the horse stalled next to your horse may not be. That means that if the horse stalled next to your horse starts a fight with your horse, and engages in biting or kicking at your horse over, under, and through the wide openings in the tubing wall, then your horse is open to injury. This potential "stall fighting" problem is the additional reason why, in my opinion, there is a need for a length of at least 12 feet of a solid, separator, kick-proof wall between horses for their protection—otherwise, where can your horse "run to" in his stall to avoid or stop being attacked?

Photo #109: Two Horses per Stall

The second problem is more subtle. I have noted a tendency for people to assume that because their horse is housed in an oversized stall that their horse does not require turnouts. That might be considered acceptable if you are able to ride your horse all the time (at least 4 to 5 times per week), but if you can't ride that often, then it really isn't the best situation for your horse. The horse's psychology, and therefore his overall well being, is greatly enhanced by giving the horse the opportunity to have turnouts where he gets the opportunity to roam more freely and naturally without having constraining tack or the demands of a rider.

Photo #110: Solid Separator Walls

The last thing that I have noted about the larger stalls is that the cowboy way of thinking is that if you have a stall that is twice as big as a minimal stall, then you should have twice as many horses in that stall. That means you could wind up having two horses in each of the stalls on either side of your horse, and that would be even more likely to inspire bouts of stall fighting.

I prefer putting my horses into a good 12′ x 12′ stall with strong and solid separator walls between each of the horses, and then paying for daily turnouts.

The next essential and vital fixture for desert barns is having automatic waterers for each of the horses. It's not an option like up north to have automatic waterers in the desert, but rather it is a necessity. Dehydration is an issue and constant threat to survival in the desert, and, therefore, the horses must have a constant supply of fresh water to drink.

Water is also a precious commodity in the desert where the limited natural sources of water have now been greatly strained by the increase in human population. The result is the need for digging deeper wells to find water, and then, even if you can find water on your land (presupposing that you can even get the rights to dig a well), you will probably find that your flow rate is unworkable. Therein lies the need for the water storage tanks which you will frequently see dotting the desert landscape outside of the towns and cities. The next problem is whether or not you will be allowed to have a water storage tank on your property which isn't universally or unilaterally allowed. Therefore, it is important to also find out if you have the right to have a water storage tank on your land before you even consider buying any land to build on.

The purpose of the water storage tank is to build up enough water pressure in the stored water so that your water demand will give you some level of minimal flow rate with which to work. That means that your well pump will run pretty much continuously to pump the well water up from the ground and into your storage tank. Then when you turn on the faucet, the pressure of the stored water and gravity will combine to give you a workable flow rate. Of course, all that pumping and storage, and even the digging of a deep well, all add cost to your construction and then to your maintenance cost of your ranch.

Photo #111: Water Storage Tank

As a side note, you'll find that the stored water tends to be lukewarm at its coldest in the summer months.

Another essential item for your desert barn is to have an insulated roof for the entire facility. Just by insulating the roof of your barns, you lower the relative temperature by tens of degrees for the stalled horses.

I have been told that local horses are able to tolerate the extreme heat of the desert summers without any further accommodation than the shade of a roof and, even better, under the shade of an insulated roof. However, my northern horses were not able to survive without further heat relieving accommodation; one of my horses suffering heat stroke that fortunately was caught and addressed before things got out of control.

What the very considerate and caring facility owners suggested beyond their box fans on each of their stalls after my horse suffered heat stroke, was the installation of a mister system. The mister system was run along the center line of the inside roofing of their stalls. The mister nozzles and tubing are hooked into a water line hose that allows cooling water vapor to be continuously spritzed down onto the horses during the high temperature, daytime hours. The cooling mist creates at least a 20 degree differential in temperature between my horses' stalls and the "outside" temperature.

Photo #112: Heat Relief Accommodations

So I would highly recommend planning both for the individual electrical outlets for each stall to accommodate box fans, as well as planning in a mister system for the entire barn. I know you're thinking that it's only because my horses are northern horses that I'm recommending a mister system. However, just think about it from a human point of view. Maybe you can tolerate a hot summer day and spend a lot of time under the shade of a tree to keep from sweating. But, given a choice on a beastly hot day, would you turn down a cooling dip in a pool plus a cooling breeze over standing under a tree and sweating? I don't think so. Are horses any different? Not really.

Another point concerning your horse's overall health about which you should be aware is that sand colic is an issue in the desert. Think about it—desert ... sand ... sand ... desert. They kind of go hand in hand, wouldn't you agree? And, in fact, they do. So,

Photo #113: Sand, Flatlands, Mountains

have desert, have sand, and then have sand with grazing horses on it? Most likely you're going to wind up having horses with sand in their guts! True, not all horses get significant amounts of sand in their guts, but if your horse is regularly foraging on the sand, you should consider giving a psyllium supplement to your horse to aid in eliminating the sand from his gut as a preventative measure if nothing else. There are two systems for administering the psyllium. One system is to provide the supplement in the horse's feed on a daily basis. The other is to do a once a month "flush" with the psyllium additive in accordance with the package directions.

Of course, the ingestion of sand can be minimized by having a rubber feeding mat placed on the stall floor or by having a hanging hay manger or hay net, as well as by installing wall mounted or hanging feed trays for the horse's grain. It is also highly recommended that you have your vet regularly check for signs of sand build up in your horse's gut.

Another big difference to be aware of with desert land is the potential for areas of swift and deep water runoff across your land from the sudden and torrential monsoon rains, the occasional winter rains, and the sudden melt off of mountain snows in the spring.

Rain on the desert tends to be an "all or nothing" event; either all the rain falls at once or nothing at all falls for days and days and even months. Such sudden deluges combined with the fact that the desert is generally surrounded by mountains where rain water tends to concentrate and then flow straight down in temporary rivers onto the flat lands of the desert, make for instant, incredibly large, deep, and fast flowing rivers—more commonly referred to as flash floods. Knowing where those flash floods will form both on and around your land is essential. If you haven't planned your barn to avoid the areas where flash floods might occur, then you could suddenly find yourself with a river running right down the middle of your barn or around the outside of your open barn, and have all your stalls flooded and your horses in a panic.

In desert towns, having flash flood rivers run across your land during a deluge is not necessarily a qualification for your land to be considered as part of a flood plain.

Depending on the circumstances of the existence of flash flood water across your land or even through your house, such criteria as for example estimating the duration of the water's presence as being temporary or brief across your property, could be considered by the town to be water that is just a "temporary nuisance." That means that going to the town hall offices to find out if your potential land purchase is in a flood plain, won't necessarily tell you everything you need to know about your land and flash flood water before you start planning where your buildings should go.

Once again, asking the locals what they know about where flash floods occur in the area and preferably where they've seen them occur on your potential land, as well as walking the land yourself and looking for telltale signs of flowing-water-disturbed land, is a good idea.

I would recommend that your barns and stalls be placed only on the highest possible land (which could literally be only a matter of inches on the flat desert land) or on mounded up land. Then creating a shallow ditch depression scraped into all the land surrounding the entire perimeter of all your buildings to form a kind of moat system along which sudden rivers of water would be encouraged and inclined to flow would be a good offensive plan. Maintaining such a "moat" would only require occasional re-dragging and is therefore a relatively easy and cost effective way to deal with flash flood water around your buildings that doesn't have an excessive flow rate or depth.

Almost all riding arenas in the desert are outdoor and uncovered. Generally only state and county fairgrounds have indoor arenas. Desert daytime temperatures are conducive to outdoor riding pretty much year round such that very few facilities even consider building any kind of covered riding ring. Personally, I would consider building a roof structure over at least one of my outdoor riding areas for protection from the intense summer heat; even a temporary, canvas type roofing seems like a good idea to me.

One last thing to consider is that having open stalls means that your horses are open not only to the air and the weather, but they are also open to the entire environment which includes the local fauna—and the desert fauna is formidable.

Having a general understanding of the likely habitats and hunting behaviors of the fauna is the best way to consider where you might want to locate your ranch in the desert. Accessing the state agricultural websites is a great way to begin a thorough investigation into what's what and what's where.

The locals have shared a few basic insights into the common fauna of the desert with me which I will share with you.

First of all, the closer you get to the mountains and the foothills of the mountains, the more likely you are to have to deal with mountain lions and bobcats and even the occasional bear. This is because the big cats don't like to wander out across open, sparsely vegetated, flat land that doesn't provide them with cover. Therefore, they tend to stay in and near their home bases in the foothills and mountains.

Next, the fewer mature trees you have on your property, the less likely a big cat will be interested in hunting on your property. Big cats like to hunt from above and drop down onto their prey, so obviously the fewer height sources you provide, the less attractive your land will be on which to hunt.

In the more open, flat land of the desert, way away from the foothills and mountains, you are more likely to have to deal with coyotes either singularly, in twosomes, and at times, in large packs of 30 or more. There are also critters called javelinas on the desert which look like wild boars, but are actually related to the South American peccaries. The javelinas are also pack animals who are generally elusive and only tend to be aggressive when they are defending their young. Of course determining whether or not you're passing by a javelina who's protecting its young is about as easy as determining when you're passing by a Grizzly bear mother who's protecting her young.

With all this predatorial wildlife existing in the desert, you clearly need to have full perimeter, protective fencing around your entire facility. But before you run off to the building supply store or brick dealer, you need to be aware that big cats and coyotes—using any kind of sturdy structure to rebound off in the process—can jump to amazing heights and therefore over very high fences and walls. For example, a mountain lion can jump straight up 20 feet and rebound jump even higher; bobcats and coyotes can jump straight up about 6 feet but also rebound jump even higher. So even if you built a strong 7 foot wall to keep out bobcats and coyotes from your ranch, you might be surprised to find a coyote or bob cat jumping up and over your 7 foot walls or fences. Coyotes, for example, have been known to jump 10 foot high fences. So clearly, trying to build a solid wall or sturdy fence around your ranch isn't going to do you much good.

Photo #114: Flimsy Wire Perimeter Fence

So what do you do? Well, the ranchers have learned that the best overall protection comes from ordinary wire mesh, five foot high fencing placed around the perimeter of your land. Because the wire fencing is cheap, both in cost and in quality, it is relatively flimsy. Because it is flimsy, the coyotes and bob cats can't use them to rebound off and scale to amplify their jumping ability. Also, the large gaps and thin wires used in the construction of the fence, are not "paw friendly" and so the animals aren't inclined to want to risk cutting their paws on the wire in a climb or an attempted rebound.

So five foot high, wire fencing, a couple of big, barking farm dogs, and a loaded shotgun or rifle is your best bet for wild animal control in the desert!

Photo #115: Major Wash in the City

Also be aware that the major washes and their branches that exist everywhere both inside and outside the city limits in the desert not only serve as pathways for the temporary raging flood waters along the land, but also provide a perfect "road" network for all the desert animals to follow. So you should realize that the closer you are to a major wash, the more likely you will be to see, hear, and potentially have to engage with the local wildlife. Case in point, even though we live in a condo in the middle of the city, we also have a branch of a major wash that runs right behind our condo. One sunny afternoon I looked out through the slider window on the backside of the condo to see a bobcat casually strolling along the top of our five foot high, cement block wall with a lunch pack of dead ground squirrel dangling from his mouth!

There are lots of other creepy, crawly creatures on the desert from lizards to tarantulas to rats, and even ugly, though important, flying things—namely bats. But the other thing to be more focused on being concerned about is the rattlesnakes.

You already know that rattlesnakes are poisonous, but what you might not know is that they are also territorial. That means that if a rattlesnake has decided to take up residence on your horse property, you must immediately get rid of it because it's not going to voluntarily go anywhere. In fact, it's going to stand its ground. Shooting rattlesnakes on sight is the cowboy standard for dealing with rattlesnakes.

Rattlesnakes are another reason why open barns work so well in the desert. What happens in the winter months is that rattlers look for a warm shelter in which to spend their winter months, and barns are their #1 shelter of choice. To have an enclosed barn on the desert, is to invite in rattlesnakes for the winter. Having rattlesnakes in your horse barn, wouldn't be good.

That's about all the information about the southwest that I have to share. Please understand again that this is just a brief overview of some of the differences in desert horse ownership and facilities. Once again there are lots of variables and opinions, but hopefully these personal experiences and observations of mine will give you a good handle on your thinking and planning for your desert horse ranch.

Lessons

Lesson #1 .... page 17 .... building restrictions

Lesson #2 .... page 20 .... perking land

Lesson #3 .... page 22 .... number of horses

Lesson #4 .... page 24 .... property equivalents

Lesson #5 .... page 25 .... existence of easements

Lesson #6 .... page 31 .... easement - legal documents

Lesson #7 .... page 32 .... land too far out

Lesson #8 .... page 32 .... wildlife

Lesson #9 .... page 33 .... neighbors

Lesson #10 .... page 53 .... survey property

Lesson #11 .... page 54 .... land purchase contingencies

Lesson #12 .... page 58 .... civil engineer

Lesson #13 .... page 60 .... building permit

Lesson #14 .... page 61 .... excavators

Lesson #15 .... page 65 .... driveways

Lesson #16 .... page 66 .... excavator - contract and schedule

Lesson #17 .... page 67 .... drainage pond

Lesson #18 .... page 68 .... firm prices and specifications

Lesson #19 .... page 71 .... house to barn utilities

Lesson #20 .... page 78 .... toxic growth

Lesson #21	....	page 81	....	landscapers
Lesson #22	....	page 82	....	excavators and rocks
Lesson #23	....	page 83	....	rocks versus stones
Lesson #24	....	page 83	....	rock size
Lesson #25	....	page 84	....	two years – pasture grass
Lesson #26	....	page 87	....	fences
Lesson #27	....	page 90	....	fence quotations
Lesson #28	....	page 98	....	barn designers vs. architects
Lesson #29	....	page 100	....	barn builder vs. house builder
Lesson #30	....	page 100	....	barn building - unregulated
Lesson #31	....	page 102	....	pole barn quote
Lesson #32	....	page 103	....	barn specifications
Lesson #33	....	page 105	....	second barn builder
Lesson #34	....	page 106	....	house builder
Lesson #35	....	page 106	....	construction loan
Lesson #36	....	page 108	....	builder references
Lesson #37	....	page 113	....	reputable bank
Lesson #38	....	page 113	....	weekly accounting
Lesson #39	....	page 115	....	hay barn
Lesson #40	....	page 117	....	separate equipment storage
Lesson #41	....	page 119	....	hay barn entry door
Lesson #42	....	page 120	....	hay barn – cement floor

Lesson #43	....	page 120	....	downspouts
Lesson #44	....	page 123	....	hay farmer
Lesson #45	....	page 128	....	commercial facility
Lesson #46	....	page 128	....	hired help
Lesson #47	....	page 130	....	commercial facility and code
Lesson #48	....	page 131	....	commercial facility and profits
Lesson #49	....	page 137	....	dumpster
Lesson #50	....	page 157	....	farriers and thrush
Lesson #51	....	page 157	....	stall mats
Lesson #52	....	page 182	....	lightning rod
Lesson #53	....	page 205	....	metal roofing
Lesson #54	....	page 271	....	short bed truck
Lesson #55	....	page 271	....	plastic truck bed
Lesson #56	....	page 272	....	tractor lifting - weight & height

PHOTOGRAPHS

Photo # 1 ... Lush Green Pasture at the End of a Late Summer Day ... Page # 7

Photo # 2 ... Our Land at First with Just the Construction Driveway ... Page # 43

Photo # 3 ... Overall Flatness of Our Land ... Page # 43

Photo # 4 ... Big Tree Spade Truck ... Page # 44

Photo # 5 ... Big Tree Planting Norway Maple ... Page # 44

Photo # 6 ... Norway Maple Tree in Pasture ... Page # 45

Photo # 7 ... Drainage Pond – First Dug ... Page # 46

Photo # 8 ... Drainage Pond – Mature ... Page # 46

Photo # 9 ... Emma, the "Old Gray Mare" -- Maybe ... Page # 50

Photo # 10 ... Aerial Photo of Farm Showing Driveways and Pathways, ... Page # 52
and Angled Corners of Pasture Fencing,
and Full Perimeter Fencing

Photo # 11 ... Buried Liquid Propane Tank ... Page # 75

Photo # 12 ... Unknown Mish Mash of Weed Growth ... Page # 78

Photo # 13 ... Prepared Pastureland for Seeding – All Natural Growth ... Page # 79
and Rocks Removed

Photo # 14 ... Derocked, Seeded Pastureland after Three Years Growth, ... Page # 85
Protective Fencing Around Pasture Tree,
& Pasture Automatic Waterer

Photo # 15 ... Fencing Protecting Pastureland ... Page # 86

Photo # 16 ... Fencing Protecting Pastureland ... Page # 87

Photo # 17 ... Why You Need to Protect Your Pastureland ... Page # 87

Photo # 18 ... Pasture Fence Hot Wire Run Inside Top Rail of Flex Fence ... Page # 89

Photo # 19 ... Horse/Man Pasture Entry Gate and Tractor Gate ... Page # 93

Photo # 20 ... Secure Gate Latch Closure Hardware ... Page # 94

Photo # 21 ... Ian and Emma Peacefully Grazing ... Page # 96

Photo # 22 ... Stage One of Our Barn Construction ... Page # 97

Photo # 23 ... Wood Carved Farm Sign with Copyrighted Farm Icon Logo.. Page #104
Photo # 24 ... Hay Barn Early Construction -- "Poles" ... Page #115
Photo # 25 ... Hay Barn Construction ... Page #116
Photo # 26 ... Hay Barn Construction Progresses ... Page #116
Photo # 27 ... Hay Barn Completed – Note Crossbucks on Slider Doors ... Page #117
Photo # 28 ... Farm Equipment in Hay Barn, Fluorescent Lights on Trusses, and Hay Bales Stacked on Pallets ... Page #117
Photo # 29 ... Muck Rack ... Page #138
Photo # 30 ... Truck Delivery of Prefabricated Arena Trusses ... Page #144
Photo # 31 ... Horse Barn Aisle Freeze Proof Hydrant ... Page #148
Photo # 32 ... Outdoor Freeze Proof Hydrant ... Page #148
Photo # 33 ... European Stall Fronts with Full Bar Grill, Swing Doors ... Page #161
Photo # 34 ... Open Stall Window with Screen and Closed Barn Grill ... Page #165
Photo # 35 ... Closed Stall Window ... Page #166
Photo # 36 ... Open Stall Window with Open Barn Grill and Screen ... Page #166
Photo # 37 ... Stall Window Exterior Latch on Wood Block ... Page #167
Photo # 38 ... Front Entry "Dutch Door" to Main Horse Barn Foyer ... Page #167
Photo # 39 ... Grooming Stall and Stall Ceiling Lights ... Page #168
Photo # 40 ... Wash Stall with Shelves, and Cross Ties with Sisal Cord ... Page #171
Photo # 41 ... Tack Room with Bridle and Blanket Racks To the Left of Entry Door ... Page #172
Photo # 42 ... Tack Room Left Wall with Wire Shelves, Blanket Racks, Saddle Stands ... Page #173
Photo # 43 ... Tack Room Back Wall with Armoire and Right Wall with Saddle Stand ... Page #173
Photo # 44 ... Barn Bathroom ... Page #174
Photo # 45 ... Viewing Room and Connector Aisle as Seen from Arena ... Page #177
Photo # 46 ... Covered Entry Porch to Horse Barn Office/Lounge/Viewing Area ... Page #177

Photo # 47 ... Lounge Area Utility Sink in Nook Area, and Aisle Entry Door ... Page #178

Photo # 48 ... Lounge Area Nook with Washer and Drier Hookups, and Utility Sink ... Page #178

Photo # 49 ... Lounge Area Exterior Wall Entry Door, and Utility Room Door ... Page #179

Photo # 50 ... Lounge Area Exterior Wall and Exterior Wall Entry Door ... Page #179

Photo # 51 ... Lounge Area with Viewing Area Cut Out, AC & Exterior Wall Window ... Page #179

Photo # 52 ... Lounge Area to Left of Aisle Entry Door showing Stairs to Viewing Area and Open Wall ... Page #180

Photo # 53 ... Lounge Area Aisle Entry Door and Stairs to Viewing Area ... Page #180

Photo # 54 ... Horse Barn Entrance Foyer ... Page #181

Photo # 55 ... Hay Barn with Cupola, and Architectural Asphalt Roof Shingles ... Page #182

Photo # 56 ... Landscape Photo of Horse Facility, and Relative Roof Pitches/Lines ... Page #184

Photo # 57 ... Bagged Shavings Storage Area ... Page #188

Photo # 58 ... Short Term, In-Barn, Hay Storage Building ... Page #194

Photo # 59 ... Grain Storage Bins in Horse Barn Foyer ... Page #196

Photo # 60 ... Grain Storage Bin with Open Lid, Hydraulic Lift, and Cable Attachment ... Page #196

Photo # 61 ... Open Storage Area ... Page #198

Photo # 62 ... Horse Barn Front Porch ... Page #199

Photo # 63 ... Horse Barn Front Porch Side View ... Page #199

Photo # 64 ... Main Barn Aisle, Suspended Lights, Eave Ceiling, Glass End Doors ... Page #200

Photo # 65 ... Porch Attic Interior Wall Vents, Stall Separator Walls, Stall Ceiling Lights ... Page #201

Photo # 66 ... Construction of Our Arena ... Page #212

Photo # 67 ... Arena Interior, Truss Ceiling, Low Bay Lights, Translucent Panels, Kick Wall, And Open Slider Doors with Kick Wall Swing Gate ... Page #214

Photo # 68 ... Relative Roof Pitches of Barn versus Arena ... Page #214

Photo # 69 ... Exterior View, Arena Wall Translucent Panels ... Page #215

Photo # 70 ... Arena Entry Gate from Horse Barn Connector Aisle ... Page #215

Photo # 71 ... Arena Slider End Door ... Page #216

Photo # 72 ... Arena Kick Wall with Cap ... Page #216

Photo # 73 ... Opened Arena Slider Showing Kick Wall Swing Gates ... Page #217

Photo # 74 ... Alex enjoying the View Thru the Opened Arena End Door ... Page #217

Photo # 75 ... Connector Aisle with Alex at Arena Entry Gate, Connector Aisle Truss Ceiling, Translucent Panels on Exterior Wall, and Spray Foam Insulation covering Crossover Heating Tubes in Aisle Bridge ... Page #218

Photo # 76 ... LeTigre Standing on Arena Entry Gate Hardware Protective Wood ... Page #222

Photo # 77 ... Angled View of Arena Entry Gate Hardware Protective Wood & Bungee ... Page #222

Photo # 78 ... Slider Door Locks ... Page #223

Photo # 79 ... Slider Door Lock ... Page #223

Photo # 80 ... Gator Pulling Arena Rake ... Page #228

Photo # 81 ... Hay Barn Metal Slider Door Ground Stabilizer (Metal Capped Post) ... Page #234

Photo # 82 ... Barn Aisle Glass End Doors ... Page #235

Photo # 83 ... Barn and Enclosed Rooms In Floor Heating Tubes Before Aisle Cement is Poured ... Page #238

Photo # 84 ... Automatic Stall Waterer and Matching Stainless Steel Feed Bowl ... Page #240

Photo # 85 ... Spray Foam Insulation ... Page #244
Photo # 86 ... Spray Foam Insulation on Exterior Wall of Barn Office ... Page #244
Photo # 87 ... Spray Foam Insulation on Aisle Wall of Barn Office ... Page #245
Photo # 88 ... Water Consumption Meter and Stall Electrical Outlet ... Page #248
Photo # 89 ... Suspended Barn Aisle Fluorescent Lighting, Barn Aisle Eave Ceiling, Box Fans on Stalls ... Page #251
Photo # 90 ... Hot Wire Junction box on Fence near Hay Barn ... Page #253
Photo # 91 ... Driveway Column ... Page #253
Photo # 92 ... Driveway Columns with Gate ... Page #254
Photo # 93 ... Exterior Light Fixtures Mounted on Wood Block ... Page #257
Photo # 94 ... Barn Aisle Rope Lighting ... Page #259
Photo # 95 ... Protective Fencing Around Pasture Tree ... Page #261
Photo # 96 ... Goose Neck, Slant Load, Three Horse, Aluminum Trailer & Tack Room ... Page #263
Photo # 97 ... Goose Neck Connector ... Page #264
Photo # 98 ... Full Rear View of Open Horse Trailer showing Slant Load with Dividers and Separate Loading Ramp ... Page #264
Photo # 99 ... Separate Loading Ramp in "Up" Position ... Page #265
Photo #100 ... Separate Loading Ramp in "Down" Position ... Page #265
Photo #101 ... Separate Loading Ramp in "Down" Position with Opened Rear Doors ... Page #265
Photo #102 ... Custom Third Restraining Wall with Locking System ... Page #266
Photo #103 ... Saddle Rack Installed in Trailer Tack Room Versus Rear of Horse Trailer ... Page #266
Photo #104 ... Battery Operated, Electro-Hydraulic Trailer Lift ... Page #267
Photo #105 ... Horse Trailer showing One Opened Window with Grill in Place ... Page #268
Photo #106 ... Horse Trailer showing One Opened Window plus Lowered Window Grill ... Page #268

Photo #107 ... 12' x 12' Open Stall ... Page #283

Photo #108 ... 12' x 24' Open Stall ... Page #284

Photo #109 ... Two Horses per Stall ... Page #285

Photo #110 ... Solid Separator Walls ... Page #285

Photo #111 ... Water Storage Tank ... Page #286

Photo #112 ... Heat Relief Accommodations ... Page #287

Photo #113 ... Sand, Flatlands, Mountains ... Page #288

Photo #114 ... Flimsy Wire Perimeter Fence ... Page #290

Photo #115 ... Major Wash in the City ... Page #291

Diagrams

Diagram #1 -- Pasture Fence Corners at 90 Degree Angles (NO!) and 45 Degree Angles (YES!) -- Page #92

Diagram #2 -- Hay Barn Diagram - 36′ x 36′ -- Page #118

Diagram #3 -- Barn and Arena Diagram with Connector Aisle and Office -- Page #146

Diagram #4 -- Arena Lighting Pattern - 4 Separate Circuits for 28 Lights -- 7 Lights per Circuit -- Page #256

INDEX

Address 60
Aerial map 54
Aerial photo
  farm 52
Air flow 159, 162, 199, 201, 203, 208
  arena 216, 220
  barn 199
  heat loss impact 242
  horse trailer 263
  pallets under hay 120
Architect 8, 98
Arena 143
  barn connector aisle 217
  design 212
  end door size 224
  end doors 215
  entry doors 223
  entry gate 215
  Eurosprinkler automatic watering 229
  exterior man door 225
  exterior wall height 213
  floor plan 146
  footing 225
  kick wall 216
  kick wall gates 217
  lighting 254
  lighting calculator 255
  lighting circuits 256
  rake or drag 228
  roof ridge vent 215
  size considerations 144
  sizes 212
  translucent panels in side walls 215
  trusses 143, 212
  turnouts 213
Attorney 18
Bank
  construction loan officer 113
  loan draw requests 113
Barn
  air conditioning 178
  air flow 199
  aisle floors 208
  aisle lights 251
  aisle rope lighing 259
  aisle walls 210
  aisle width 161, 208
  aisles 208
  arena connector aisle 217, 232
  bathroom 142, 174
  bathrooms, number of 19, 20
  builder quotes 101
  builders and finishers 100
  cat 192
  ceiling fans 201
  cement aisle floors with brush finish 155
  colors 103
  doors, end 235
  dust 195
  electrical outlet boxes 248
  electrical wiring in metal conduit 248
  electricity 246
  end door locks 223
  entrance foyer 142, 181
  exterior electrical outlet boxes 252
  exterior lighting 257
  exterior motion sensitive lights 258
  exterior wall height 199
  floor plan 146
  flow between areas 144
  foyer chandelier 251
  freeze proof water hydrants 147
  grooming stall 141
  hay storage shed 194
  heat in the aisles 241
  heating system 243
  heating system in the floors 237
  lighting - tack room, bathroom, office 251
  lighting fixtures 249
  location 47
  location of light switches 249
  man door entrance 167
  number of stalls 139
  office/lounge/arena viewing room 175
  office/lounge/kitchen & arena viewing area 142
  plumbing 237
  rafter vs. truss ceiling 204
  second builder, finisher 105
  spray foam insulation 244
  stall automatic waterers 147
  stall floors 155
  stall lights 250
  stall waterers and feeders 239
  storage areas 142, 185, 191
  tack room 142
  telecommunications wiring 260
  utility room 142, 177, 178
  utility sink in the office 246
  wall separation with arena 202
  wash stall 142
Barn aisles
  arena connector gate 219
  heated 241
Bathroom
  number of 19, 20
Blanket
  storage 174
Broadband
  access and service 76
  Internet access options 77
Builder
  barn vs. house 100

Building
elevations 47
inspector 110
Building permit 60
Cat
barn 192
Commercial facility 128
Connector aisle
arena entry gate 218
exterior door 219
exterior wall height 218
garage door into arena 219
translucent light panels 218
Construction driveway 53
Construction loan 106, 113
your approval for draws 113
Culverts
driveway 62
Cupola 181
Dogs
farm 193
Doors
barn entry 167
Driveways
construction 53
culverts 62
dust control - calcium choride 232
gravel vs. asphalt 62
truck turnarounds 53
turnaround loop 53
Dust
barn 195
Easements
access 17
property 25
utilities 25
Electricity 71
backup generator 72
distribution within farm 72
pvc pipe under driveway 65, 72
Electricity/electrical
400 amp service 72
generator 247
power and circuit distribution 247
Engineer
septic system 20
Eurosprinkler 229
Excavator 61
contract 66
phases 68
trenches for utitlities 69
Farm
entry gate and electrical circuits 253
equipment 271
name 104
Feed bowls
stall 239
Fencing 88
45 degree corners 90
full perimeter 94
hot wire circuit 253
location and type 50
Fire protection 23
Flies 134
Footing
arena 226
arena moisture 229
Foyer 181
Generator 72
transfer switch 72
Grain
bulk 196
storage in the barn 195
Grandfather clauses 24
Grass
pasture mix for horses 80
Hay
amount horse eats per day 119, 126
barn 115
barn - lighting 259
barn, cement floor 120
barn, downspouts 120
barn, entry doors 119
barn, floor 118
barn, man entry door 119
barn, roof pitch 118
barn, size 118
cuttings 125
farmer & supplier 123
nutritional analysis 124
storage capacity in barn 118
Hired help 128
Horse
restrictions on number of 22
trailer 263
Horses
worming 197
Hunting season 38
Insulation
spray in foam 244
Insurance
liability umbrella 34
title 29
Land 13
aerial map 54
attorney 18
envisioning location of buildings and other things for a site plan 41
farm land rezoned for taxes 23
lay of land 47
location 31, 40
low spot 45
perked 18, 20
plant/weed growth 78

purchase contract contingencies - septic 22
purchase statute of limitations.................................... 31
rough sketch of a site plan.. 42
septic system.. 18
soil distribution 48
walking 54
well 18
Landscaper
commercial for pastures.. 82
Medical supplies
basic horse 39
Muck
rack 137
trash removal.. 136
Natural gas.. 73
Neighbors.. 33
bobblehead.. 37
dogs.. 33
walking through the woods...................................... 36
Office/lounge/kitchen/arena viewing room.......... 175
Paddocks
run-in 95
Pasture
gates - two sizes... 93
grass cutting height... 85
Pastures.. 48
fencing type and installation time 87
grasses.. 78
removing rocks 78
run-in shelters 262
separation.. 48, 93
Perk
land 20
Perked.. 18
Police protection 23
Pond 45
drainage... 67
Propane... 73
Rat poison
hay barn 192
Restrictions
building.. 17
number of horses... 22
Riding ring
outdoor.. 226
Road system.. 62
Rock removal
commercial equipment.. 82
Roof
asphalt shingles.. 207
metal 205
ridge vent.. 208
snow load.. 208
Roof materials.. 204
Roof pitch
arena.. 143
hay barn... 118
horse barn.. 200
Run-in
shelter 42
Run-off water 46
Salt blocks
stall... 240
Sand
seam 21
Security system 23
Septic 75
engineered system 20
field 20
field drainage pipes 21
number of baths.. 20
number of baths supported 19
permit 55
sand mound 20
sand seam 21
Shavings... 185
analysis for bagged 187
analysis for bulk 185
Site development 58
Site plan
building locations and elevations.............................. 47
civil engineer.. 58
driveways and pathways... 52
first step 41
pastures... 48
road system... 62
rough sketch 42
utilities.. 70
Soil distribution... 48
Stalls
air flow 159, 162
electrical outlet boxes... 202
fans 202
floors 155
grooming 141, 168
mats, impervious, interlocking 157
number.. 139
separator walls.. 159
shavings... 158
size 139
stall doors 162
stall fronts.. 160
wash 142, 169
windows 163
Storage
enclosed grain bins 196
grain in barn 142
grain in the barn.. 195
hay in barn.. 142
hay in the horse barn 194
hay, short term.. 191
shavings in barn.. 142
small equipment 191

Streams ... 45
Survey ... 53
Surveyors ... 53
Tack room ... 172
Tax id number ... 59, 60
Taxes
  assessment equivalents ... 24
  property - farm ... 22
Telephone
  service ... 76
Television reception
  cable or satellite ... 76
Title Insurance ... 29
Trees ... 44
  big trees ... 44, 261
Trenches
  utilities ... 74
Truck
  trailer towing ... 268
Trusses
  arena ... 143
Utilities ... 70
  trenching ... 69
Utility room ... 177
  entrance door ... 181
Walk the land ... 54
Walls
  separator - barn & arena ... 202
Wash
  stall (rack) ... 169
Water ... 71
  freeze proof hydrants ... 71, 245
Water troughs
  pasture with heaters ... 241
Waterers
  automatic in pastures ... 240
  automatic in stall ... 238
  consumption meters ... 248
  daily cleaning ... 241
  heater wiring ... 248
  heaters ... 241
Well ... 56
Wildlife ... 32
Windows
  air flow ... 163
  European stall windows ... 164
  stall ... 163
Zoning ... 14